Isau A. B. Quissindo
Freitas M. V. Cachenhe

Carbon Sequestration in Native and Exotic Forests

Isau A. B. Quissindo
Freitas M. V. Cachenhe

Carbon Sequestration in Native and Exotic Forests

A look at the forests of Huambo (Angola)

ScienciaScripts

Imprint

Any brand names and product names mentioned in this book are subject to trademark, brand or patent protection and are trademarks or registered trademarks of their respective holders. The use of brand names, product names, common names, trade names, product descriptions etc. even without a particular marking in this work is in no way to be construed to mean that such names may be regarded as unrestricted in respect of trademark and brand protection legislation and could thus be used by anyone.

Cover image: www.ingimage.com

This book is a translation from the original published under ISBN 978-620-3-46582-2.

Publisher:
Sciencia Scripts
is a trademark of
Dodo Books Indian Ocean Ltd., member of the OmniScriptum S.R.L Publishing group
str. A.Russo 15, of. 61, Chisinau-2068, Republic of Moldova Europe
Printed at: see last page
ISBN: 978-620-3-59284-9

CARBON SEQUESTRATION IN NATIVE AND EXOTIC FORESTS

A look at the forests of Huambo (Angola)

Isaú Alfredo Bernardo Quissindo "Josué

Freitas Moisés Ventura Cachenhe

TECHNICAL DATA

Title: CARBON SEQUESTURE IN NATIVE AND EXOTIC FORESTS - A look at the forests of Huambo (Angola)

Author: Isaú Alfredo Bernardo Quissindo "Josué" & Freitas Moisés Ventura Cachenhe

Cover: I. A. B. Quissindo "Josué

Copyright © 2021 by I. A. B. Quissindo & F. M. V. Cachenhe

Publisher: New Academic Editions, Germany

INDEX

LIST OF ABBREVIATIONS AND ACRONYMS

CFC	Chlorofluorocarbons
DAP	Diameter at Breast Height
DCP	Project ConcePt document
GHGs	Greenhouse Gases
HFC	Hydrofluorocarbons
IIA	Institute of Agricultural Research
IPCC	Intergovernmental Panel on Climate Change
kg	Kilograms
m3	Cubic Meter
CDM	Clean Development Mechanism
mm	Millimeters
NDVI	Normalized Difference Vegetation Index
NGOs	Non-Governmental Organizations
PFCs	Perfluorocarbons
Ph	Hydrogen Potential
ppb	Parts Per Billion
PPm	Parts Per Million
PPt	Parts Per Trillion
CERs	Certified Emission Reductions
REDD	Reducing Emissions from Deforestation and Degradation
	Forestrv

DEDICATION

To all forestry engineers and environmentalists in Angola and around the world, who day by day do not measure efforts to make the good management of natural resources, particularly forests, aiming at the common welfare.

PREFACE

In the more than 100 pages of this book, the foresters who dared to write this work present the rationale, methodology and results of a study that compared the carbon sequestration capacity of a natural forest and an exotic forest. The data and the basis of the study was the traditional forest inventory, which gives an added scientific value to the work, because it not only addresses issues of carbon storage, but also that which is one of the preponderant activities of every forest engineer.

They began by raising questions about climate change, deforestation, and anthropic actions that accelerate that process, and continued by addressing the role of forests in maintaining life on the planet, since, among other things, forests are fundamental to the planet's climate regulation.

In addition to the short introductory and background note chapters and the conclusions with corresponding recommendations, the book has three (3) large chapters.

In chapter 1, the authors make an approach to the state of the art of the subject "forest and its relationship with the environment". In this chapter approaches were made such as: iimportance, current situation and function of forests, deforestation and main problems of forests, proposed practices to mitigate deforestation at a global level, biomes of Angola, climate change in the world, greenhouse effect, carbon cycle, carbon sequestration in forest plantations, among others.

Next, the authors describe in the second chapter the "data, procedures and methodology" applied for the comparative study of carbon sequestration capacity by native and exotic forests in Huambo (Angola).

Aspects on establishment of the sampling plot, forest inventory, age estimation of the planted and miombo woodlands, species identification, determination of dasometric parameters, were presented. In addition, the authors present remote sensing data (Normalized Difference Vegetation Index) of the two forests studied, as well as the ways of estimating carbon sequestration and the main formulas.

In the third chapter, as the core of the study, results of the evaluation of the data analyzed in the previous chapter were presented. In this chapter on "carbon sequestration by planted and native forest in Huambo", the researchers present data in graphic and alphanumeric form related to forest inventory, spatial distribution of biomass and carbon, canopy cover fraction, above ground biomass, net biomass rate, annual biomass rate and carbon sequestration of the Brito Teixeira planted and Miombo native forests.

This work explains and reinforces the need to take greater care with natural resources, particularly forests due to their environmental and protective role. For this reason, I recommend to all those who work with these issues to read this work that, in an expeditious way, facilitates the introduction to the study of carbon sequestration by the main carbon sinks.

Angel de Miguel Garcia - PhD,

Wageningen Environmental and Research, Wageningen, The
Netherlands.

INTRODUCTORY NOTE AND BACKGROUND

INTRODUCTORY NOTE AND BACKGROUND

Climate change is one of the most serious environmental problems faced in recent years and can be considered one of the most serious threats to environmental sustainability, health, human well-being, and the global economy50.

Climate change is already evident, being reflected in rising air and ocean temperatures, melting snow and ice, and rising average sea levels. Some of the predicted consequences of changes in the climate system include increases in: the number of people exposed to water shortages, storm and flood damage, species extinction, and changes in the distribution of some disease vectors47.

Notoriously, the use of fossil fuels is the human activity that represents the greatest source of greenhouse gas emissions, plus changes in land use also contribute significantly to increasing the concentration of these gases in the atmosphere. The FAO considers that anthropic activities have already converted more than 27% of the earth's land surface into agro-pastoral areas, with 13 million forest hectares being deforested every year28.

The reduction in forest stocks is a barely visible, though important, phenomenon. During the past 25 years, global carbon stocks in forest biomass have decreased by almost 17.4 giga tons (gt). This reduction resulted mainly from the conversion of forests to other uses and forest degradation31.

It is estimated that deforestation and forest degradation rates in developing countries account for about 12-20% of annual GHG emissions81,95.

To understand the true function of forests and plants in the life of the human being, it is necessary to point out that an adult needs, on average, 15 m3 of air per day, compared to 2 liters of liquids and 0.5 kg of solid food. Air is an indispensable element to the human being, who can survive 5 weeks without food, 5 days without water, but the vast majority of people do not survive more than 3 minutes without air, this resource is essential to the senses of sight, smell and hearing, the first two being directly affected by air pollution90.

The forests are fundamental for the planet's climate regulation, through photosynthesis and respiration, trees recycle atmospheric carbon and capture CO_2 to release O_2, which makes them the planet's lungs, propitiating life on our planet, since that is where the entire food chain begins96.

The IPCC recognized that reducing forest destruction has an important role in reducing global carbon emissions and since then, REDD has taken center stage in the international debate on climate change. Today they represent an important short and medium term strategy to minimize the negative effects of gas emissions on the planet through the conservation of forests, whether native or planted48, 49.

Forests represent an essential solution to the problems associated with climate change and the mitigation of its effects, when sustainably managed they increase the resilience of ecosystems and societies and allow the optimal use of the role of forests and trees as carbon sinks and reservoirs while providing other environmental services32.

Trees are able to capture a significant amount of CO_2 from the atmosphere and store it in their leaves, branches, stems, bark, and roots, with the stored carbon accounting for 50% of the weight of a tree's biomass51.

The photosynthesis carried out by land plants is responsible for retaining atmospheric carbon in the plant material and eventually in the organic matter in the soil. Therefore, ecosystems with more biomass and little disturbed soil, such as forests, retain carbon in the form of CO_2 on a much longer time scale, on the order of decades and centuries63.

Angola has cyclically felt the effects of climate change; the country has faced prolonged drought, especially in the East and South, floods and wildfires. As a result, agricultural production has been threatened due to unavailability of water, reduction of terrestrial and marine biodiversity; all of this has negative implications on health and food security71.

Among the various factors associated with this may be the effects of deforestation. Deforestation rates in Angola, though little studied, have been gradually increasing over time22. Locally, for example, a study on deforestation and forest degradation has already been conducted, where local researchers consider there to be a significant decrease in the Miombo forest area in Huambo Province between 2002 and 2015, and severe forest degradation69.

However, in view of the constant deforestation rates and the lack of forest replanting currently observed in Huambo, these represent future failures in the process of carbon sequestration by the forests of Huambo Province, which in turn has implications for climate change at the local, national, regional and universal levels.

Therefore, it was considered that the estimation of the amount of carbon sequestered per unit of area or forest surface, whether planted or natural, and its relationship with the deforestation rates will serve as a basis to substantiate the need to adopt sustainable forest management in Huambo and in the country.

Since the carbon stored by native and exotic forests have been little studied in the country, this study aimed to assess the carbon stock potential sequestered by Brito Teixeira planted forests and native Miombo forests in Chianga (Huambo).

To do this, it was necessary to perform:

- A survey of forest inventory data on rectangular plots in the study area;

- The estimation of carbon sequestered by Brito Teixeira and Miombo da Chianga forests;

- The comparison of carbon sequestration capacity between an exotic forest (Brito Teixeira) and a native forest (Miombo), in order to propose conservation and preservation measures for these forests and beyond.

CHAPTER 1. THE FOREST AND ITS RELATIONSHIP WITH THE ENVIRONMENT

CHAPTER 1. THE FOREST AND ITS RELATIONSHIP WITH THE ENVIRONMENT

1.1. Concepts and generalities

In general, it is considered that there are more than 600 concepts of forest. However, FAO defines forest as an area equal to or greater than 0.5 ha with trees greater than 5 m in height and crown cover greater than 10%, or trees capable of achieving these parameters. This does not include land that is predominantly under agricultural or urban use30. From an ecological point of view, the forest is a terrestrial ecosystem organized in overlapping strata (mossy, herbaceous, shrub and/or arborescent), which allows maximum use of solar energy and greater diversification of ecological niches16.

A forest is an area of at least 0.05-1.0 ha with canopy cover (or equivalent density) of more than 10-30%, with trees having the potential to reach a minimum height of 2-5 meters at maturity *in situ.* A forest can consist of both dense forest formations, where trees of various strata cover a high proportion of the ground, and open forests. Young natural stands and all plantations that have yet to reach a density of 10-30% and a height of between 2 and 5 meters are included as forest, as are areas that are normally part of the forest area and that are temporarily deforested as a result of human intervention such as harvesting or natural causes, but for which forest reversion is expected93.

A natural forest is an arboreal ecosystem characterized by the presence of trees and bushes of multiple native species, of varying ages and heights, regenerated by natural succession, with an incredible biodiversity of plants, animals and microorganisms, living in harmony, that is, they are areas composed of native trees, not planted by man96. These, according to the UN, can be classified in dense forests (when there are trees of different heights that cover a large part of the land or more than 40% and do not have a continuous dense herbaceous extraction) and open forests (formations with a discontinuous distribution of trees, but with a canopy cover of at least 10% and density of less than 40%)[68].

According to FAO, a planted or exotic forest corresponds to an area of planted forests (composed of planted trees and/or the seeding of native or exotic/introduced species)[31]. Currently, the world has approximately 290 million hectares, corresponding to 6.6% of the planet's forest area; by 2020 this area is projected to increase to 300 million hectares32.

1.2. Importance, current status of forests

Forest resources are of great importance because they offer ecological, social, and economic benefits, and it is estimated that worldwide some 60 million people are almost completely dependent on forests for survival, and that 350 million people living near forests rely heavily on them for their livelihood and income18.

For a local researcher in a village, not everything is paid for with money, water is taken directly from the source and food comes mainly from the forest or the land and housing is built with materials gracefully cut from the forests52.

The importance of forests is seen by the thousands of people in the world's poorest countries, including indigenous groups that survive on the supply of food, building materials, water, and medicine obtained from forests. In addition, forests have cultural and spiritual value in a people96.

Forests also contribute to climate change mitigation by sequestering atmospheric carbon, both above and below ground and in the soil, and thus reducing the greenhouse effect42.

Natural outdoor areas such as forests, parks, trees and gardens are known to offer opportunities to improve public health and well-being. Contact with the natural environment can provide an antidote to some of the healthy aspects of the urban lifestyle, and there is a growing realization that this should influence the way our surroundings are planned and managed. Trees and other plants have been used in traditional, modern and alternative medicine as sources of pharmaceuticals and other chemicals65.

Data from the National Forestry Development Institute point out that as far as their protective function is concerned, forests provide habitat for almost two thirds of all species on the planet and regulate the climate42.

Despite the value of forests, according to FAO, on a global scale, natural forest area tends to decrease in relation to planted forest, which tends to increase31.

Forests cover more than 4 billion hectares, or 31% of the world's land area. It is estimated that about 290 GtC are stored in their biomass alone. But with the average annual conversion of 13 million hectares of forest to other land uses in the last decade, this stock is being reduced annually28.

In 2005, the total forest area was 3952 million ha, which represents 30% of the total area of the globe. Of this figure, 93% correspond to natural forests and 7% to forest plantations24.

Of these, about 36% of the total forest area is classified as primary, that is, naturally occurring species forests, in which there are no clearly visible indications of human activity and the ecological processes are not greatly altered. The number of forest species estimated worldwide is 100,000, most of them from tropical zones and some of them still unidentified. However, forests are being exploited uncontrollably in recent times, so that a very dangerous phenomenon of deforestation is taking place that has already resulted in the annual loss of 8.9 million total hectares between 1990 and 2000, according to FAO data25.

Africa's forest cover is estimated to cover an area of 650 million hectares, comprising 17% of the world's forests23. The main forest types are tropical dry forests in the Sahel region and eastern and southern Africa, tropical moist forests in western and central

Africa, subtropical forest and shrub formations in northern Africa and the southern tip of the continent, and mangroves in coastal areas. These forests have several international biodiversity critical zones and only 1% of Africa's forests have been planted62.

Meanwhile, annual forest loss in Africa between 2000-2005 is estimated to have been 4,000,000 hectares (55% of the total forest lost worldwide in that period). The burning of the forest zone has a very significant contribution in the emission of CO_2, other gases and aerosols into the atmosphere; fires in African savanna zones account for 22% of the forest area burned annually in the world15.

In Angola, there is a forest extension of approximately 53 million hectares (43.3% of its land area). These forest resources are a source of subsistence and income for most rural populations, especially for those in vulnerable conditions59.

The potential of Angola's forests and wildlife is an element of primary importance in the country's sustainable development process. Reinforcing this statement is the fact that forests today, by their very nature, play a leading role in strategies to combat hunger and poverty. Thus, any coherent and sustainable development strategy, along with those that integrate economic and social concerns, must be related to the conservation and use of these resources. Forests have important socio-economic, conservation and protective functions, in addition to an environmental and ecological function in the conservation of water, soil and biodiversity42.

As for basic forest resources, Angola has 2.669.700 hectares that integrate 18 forest reserves, created before 1975, as strategic reserves for future extraction of raw material60 . The objectives of the creation of these reserves are: conservation (forests, rare species or of scientific value, soils, etc.), regularization of hydrographic and climatic regimes43.

In Angola, Miombo woodlands occur in the Zambeziaco biome38, [78,] which covers more than 80% of the country's area and includes a variety of forests, where the main one is the Miombo woodland57. These forests are generally composed of species of the genus *Brachystegia, Julbernardia* and *Isoberlinia1*[38], *Erythrophleum, Uapaca* y *Tetrorchidium9.*

Recent data show that by the year 1980 the country had 148,000 hectares of forest plantations, and 59% of these forest plantations (86% *Eucalyptus spp,* 10% *Pinus patula* and 4% *Cupressus spp.*) are located in the Central Highlands of Angola, between Bié, Huambo, Huila and Benguela29, [59, 60].

1.2.1 Function of forests

According to FAO, forests can perform different functions such as31:

■ **Productive function**:

Forests are suppliers of many products, both wood-based products (NTFPs) as well as non-timber forest products (NTFPs). To fulfill this production function, approximately half of the world's forests are designated for production. Likewise, a high percentage of forest plantations are undertaken for this purpose, i.e., for production. In fact, currently about 109 million hectares of forest plantations are carried out with this production function and this rate is tending to increase; the areas of forest plantations for production increased to 2.5 million hectares annually in the period 2000-2005. The world wood extractions in 2005 amounted to 2800 million m^3 of which 40% was for firewood (120 million m^3). As for the extractions of non-timber forest products, the data presented by FAO in its report on the evaluation of forest resources31 are not very consistent, although it would appear that in the period 1990-2005, there seemed to be a trend towards an increase in the extraction of these products. Given the economic data of the sector, we can cite some interesting examples to focus the importance of forest resources worldwide. The forestry sector provides employment for 14 million people worldwide. The world's gross value added in the forest sector is US$ 468 billion. World trade in wood products was $235 billion in the year 2008. World production of timber was 3449 million cubic meters, also in the year 2008. Forests also play an important role in energy consumption; more than 2 billion people depend on wood fuel for cooking, heating their homes, food preservation, and other purposes. In fact, wood fuels account for 60% of the world's consumption of forest products. Biomass energy accounts for 15% of the world's energy consumption, rising to 90% in some developing countries31.

■ Protective function

Forests provide habitat for almost two-thirds of all species on the planet, so ensuring good forest condition guarantees biodiversity, soil and water conservation, water supply, and climate regulation. Forests influence the climate because they reduce the level of heat reflected into the atmosphere compared to other, less vegetated ecosystems. In addition, the trees that make up the forests mitigate sudden changes in temperature; on the one hand, in hot weather they provide shade and absorb heat, producing a cooling effect, and in hot weather they stop, filter, and divert wind, reducing the feeling of cold. They also protect against wind erosion, especially in coastal forests, reduce sedimentation on the coast and the impacts of storms and tsunamis, protect against land movement caused by intense atmospheric phenomena such as a *tsunami, and* filter air contamination by retaining airborne particles. In addition, they have functions to conserve biological diversity or natural or cultural features. Another of the most important protective functions to highlight in the current situation of climate change, is the one developed through the carbon cycle. Biomass and dead wood absolve respectively 44% to 6% of the total carbon in the forest ecosystem, while soils up to 30cm of surface plant matter contain around 46% to 4% respectively. Worldwide, forests designated for protection as a primary function occupied in 2005, 348 million hectares, corresponding to 9% of the world's forest area, and as a protective function, although not necessarily as a primary function, had 1190 million hectares of forests.

■ Socioeconomic function

The socioeconomic functions of forests are the most difficult to quantify and vary depending on the degree of development of each country. In post-industrial developed societies, the recreational benefits of forests and their value as places for recreation or for maintaining a rural way of life can be very important, while in developing countries the importance of the forest is based on the number of jobs it generates, or the type of activities that take place. The resources obtained from forests are very important in both the food and pharmaceutical sectors. As a complement to food, food products from the forests provide an indispensable nutritional support, and can avoid situations of permanent hunger in situations of drought, floods, or pests and diseases of non-forest crops. In the pharmaceutical sector, about a billion people around the world depend on medicines derived from forest plants to cover their medicinal needs. In addition, populations in developing countries rely on up to 75-90% of natural products as their sole source of medicines27.

1.2.2. Deforestation and major forest problems

Today, deforestation is a major global concern due to humanity's colossal capacity to exploit and devastate forest resources. Today we speak of the extinction of most of the world's primary forests; data from GREENPEACE reveal that Western Europe has already lost 99.7% of its natural forest cover; Asia, 94%; Africa, 92%; Oceania, 78%; North America, 66; and South America, 54%.This alarming loss of primary forests, among other factors, was due to the historical demand for land, wood products and energy, especially during the 20th century, and also to natural causes39, [10].

The main problem with forests today is deforestation and it can be caused by human or natural factors. The former are much more frequent than the latter, since deforestation occurs when people eliminate forests and use them for other purposes, such as agriculture, infrastructure, human settlements. Natural phenomena, specifically disasters, can lead to the conversion of forests to other land uses32.

Among others, deforestation results in increased soil erosion and a decrease in biodiversity, the silting up of river systems, loss of soil, a general loss of biomass, as well as a reduction in the capacity of forests to sequester carbon, all of which bring with them worrying social, economic, and environmental consequences that in many cases are difficult to combat44.

Forests are fundamental to the protection of essential ecosystems and related services in all provinces of Angola, although forest cover and composition varies greatly from province to province61.

1.2.3. Drivers and consequences of deforestation

There is a very broad consensus that the various forms of or changes in land use - often for obtaining raw materials - are the main factor in the disappearance of forests. But among several factors, logging, the transformation of forests into arable farmland, pastures, forest fires and the use of firewood as fuel are determining factors of deforestation in the world26.

Even if it is impossible to foresee all the consequences, it is clear that deforestation directly interferes with fauna, devastates flora species, contributes to water and air pollution, increases acid rain and the greenhouse effect, affects the local climate and even the planet as a whole, increases the ease of soil erosion and decreases biodiversity, siltation of river systems, loss of soil, general loss of biomass, and reduces the capacity of forests to sequester carbon61.

1.2.4. Implications of deforestation on the hydrological regime

According to recent data, forest cover has a favorable influence on soil hydrology, insofar as it facilitates infiltration, percolation, and water storage processes, reducing surface water runoff and, consequently, soil erosion. The effects of deforestation on the hydrological regime translate into89:

- Reduction of relative humidity

The transpiration of the leaves, besides regulating the humidity of the air, also functions as a regulator of the temperature of the environment. Deforestation, in this case, leaves the air drier and the temperature higher and more inconstant. Forests increase both the abundance and frequency of local precipitation compared to open areas.

- Water quality problem

Erosion and leaching are two direct causes of deforestation that cause water to lose quality, making it more turbid and, in certain cases, unfit for human consumption.

- Soil erosion

Modern man has perfected two techniques capable of destroying mankind: total war, the war of the universe, and worldwide soil erosion. Of the two, the most insidious and fatally destructive is undoubtedly erosion. Vegetation, understood as forest cover, plays a preponderant and decisive role in attenuating the impact of raindrops, whose impact on the soil surface can initiate the erosion process. Erosion is a process of separation of soil particles carried out by water and/or wind, which is sometimes the indirect consequence of various actions carried out mainly by man, through the practice of burning, cutting down vegetation, grazing. This means that an unprotected soil is vulnerable to the intensive and frequent process of erosion.

1.2.5. Proposed practices to mitigate deforestation globally

According to FAO the responses to deforestation can be considered under several categories such as27:

- Attempts to stop deforestation through protective legislation, such as the creation of Forest Reserves, with full protection or limited and controlled forms of forest exploitation. For these measures to be successful, however, they must first take into account the causal factors that are responsible for deforestation.

- Rehabilitation of degraded or cut forest through replanting. This can be achieved by policy reforms aimed at improving land management. Forest policy reform, strengthening land use rights or extending commercial forest concessions to promote replanting, is a priority in many cases.

- Planting of new forest areas to partially compensate for the deforestation of other areas. This course of action, like the previous one (rehabilitation / replanting) will, however, be limited to the creation of planted forests which are usually composed of a single species or a small number of them. As such, they can in no way be expected to compensate for the high levels of biodiversity found in most natural forests in both tropical and temperate zones.

- Use of forests as a source of Carbon storage.

- Measures to provide alternatives to the products obtained by deforestation. For this approach to be successful there needs to be an understanding of the social, environmental and economic factors involved in creating pressure on the forest.

- Implementation of social, environmental, and economic studies to determine the causal factors involved in deforestation, and creation or strengthening of institutions with the authority to verify these factors.

- Development of communication mechanisms within and between government departments, and between them and NGOs to facilitate the transfer of information from these studies into policy action.

1.3. Biomes of Angola

Angola's phytogeographical chart mentions 7 national biomes and they are defined according to their floristic composition and geographical location below the description of each one5:

- Dense humid forest with high rainfall. It corresponds to about 2% of the total forested area, covering the rugged terrain of the Atlantic coast, from Cabinda to the Balombo River, with strong expression in Alto Maiombe (north of Cabinda) and in the Dembos (triangle formed by the provinces of Uige, Bengo and Kwanza Norte). This presents a very varied floristic composition, with several higher arboreal strata.

    o Several species of *Albizia, Celtis, Ficus, Pycnantus angolensis* stand out.

- Guinean Savanna Mosaic. It occupies about 20% of the total forested area, characterizing the septentrional part and part of the humid zone of the Angolan territory.

- In this type there is a domain of Guinean-type savanna, with scattered trees and shrubs, most often of *Hymen oca rd ia acida, Maprouena africana, Nauclealatifolia, Annonarenaria,* and the *Piliostigmathonningii,* alternating with gallery forest.

- Open Miombo woodland. This type of forest occupies about 45.2% of the total forest area and is spread over vast areas of the country, including the provinces of Huila, Cuando Cubango, Moxico, Bié, Huambo, Malanje, Benguela and Cuanza Sul. It presents several dominant associations, where the most frequent are the genera *Isoberlinea, Brachystegia* and *Julbernardia*. It is of medium productivity in terms of commercial wood. However, it has a high use value in terms of woody fuel, construction materials, food products and medicinal plants.

- Dry Savanna with trees and/or bushes. It occupies about 24.2% of the total forest area, extending mainly in the coastal provinces and some in the interior. It is a forest type with low productivity in terms of commercial timber, but holds high social value in terms of woody fuel, construction materials, food products, and medicinal plants.

- Grasslands, Chanas and anharas. It consists of herbaceous formations of the floodable surfaces as well as mangroves. They are of low productivity in almost all aspects and occupy about 5.3% of the total forest area.

- Steppes of the sub-desert coastal belt. They occupy about 3% of the total forest area and are of low productivity in almost every respect.

- Desert vegetation formations. This type occupies about 0.3% of the total forest area and is of low productivity in all aspects.

1.4. Reducing biodiversity and preserving biological diversity

For some authors, the concern for preserving forests should be in first place because their loss would impoverish the Earth's biodiversity. Because the forest ecosystem, in this context, is considered a reservoir or habitat par excellence of biological diversity and its destruction implies the disappearance of several species of plants and animals found there[85].

The reasons for preserving biological diversity are as follows[78]:

- Because all living things play a decisive role in the global cycle of matter, of the climate, and of all renewable resources, without which the existence of humanity is inconceivable;

- Because all biological species are essential from an economic point of view to the extent that half of the world economy depends fundamentally on the use of wild species in agriculture, medicine and industry; even tourism in its leisure component would be impossible without living nature;

- Because from an aesthetic point of view biodiversity is of immeasurable importance and value;

- Finally, the World Charter for Nature grants the right to exist to all species regardless of their importance to man, i.e. all components of biodiversity participate in the universal processes of production, maintenance, and regulation of life; the loss or degradation of biodiversity can thus have economic, social, and cultural impacts and costs, in addition to having profound ecological, ethical, and aesthetic implications[10].

1.5. Climate change in the world and the Kyoto Protocol

It should be noted that vulnerability to climate change also depends on the social characteristics of the region. There is a particular concern with developing countries like China or India, where social vulnerability, poverty and inequality intensify existing climate problems[80].

Before the Industrial Revolution until the mid-twentieth century, countries believed that economic growth was synonymous with the exploitation of natural resources; moreover, it was thought that the environment was only a source for the extraction of raw materials; they had no concern for the environment; on the contrary, exploitation was the main way to acquire wealth[14].

The Industrial Revolution spread rapidly throughout the world, drastically and perpetually changing our way of living, of seeing the world, of experiencing it, and of disposing of its resources. In this way, mechanized production brought benefits, but also a generalized increase in the consumption of natural resources[53].

After this period began a relentless search for energy for industries to maintain and improve their production process. The main resources extracted from the environment to supply energy for these production processes were fossil fuels. First, the burning of coal was used, then, with technological advances, there was a change to oil, and later to natural gas.

This industrial process is largely responsible for the unbalances in the biogeochemical cycles of carbon, nitrogen and sulfur in the various ecosystems. The burning of these fuels releases excessive amounts of these gases into the atmosphere, which causes an abnormal warming of the planet. Thus, climate changes began to occur at a much faster rate than if they occurred naturally and this is the main problem of these changes in the climate, the warming of the planet[84].

The intensification of the greenhouse effect is directly linked to human activities, mainly from the burning of fossil fuels. This burning occurs in domestic and commercial uses, in transportation activities, in power generation, in industry and in agriculture[37].

Thus, man since the beginning of the 21st century is polluting the environment more with industrial processes, agricultural activities, inadequate waste treatment and with the increase of deforestation. Greenhouse Gases are the biggest sources responsible for global warming[64].

The consequences of this global warming are: melting of the polar ice caps, which causes coastal areas to flood; elimination of biodiversity from the planet; increased desertification of areas; increased frequency of droughts and floods; reduced crop yields; damage to the health of the population due to heat waves; increased occurrence of hurricanes and cyclones; and the spread of contagious diseases53.

In this way, the problem of climate change is directly linked to the energy options adopted by the rulers of each country, in addition to people's consumption patterns. This relationship between those who decide for what and the others is a cruel relationship, because all countries will suffer from the environmental, economic and social impacts caused by the implementation of an energy matrix based on fossil fuels84.

In 1988 the testimony of NASA physicist James Edward Hansen pointed to scientific evidence that humans were interfering with the climate, and this sparked an awareness of global warming91.

Also in 1988 in Toronto, during the Conference on Atmospheric Change, the Intergovernmental Panel *on* Climate Change (IPCC) was created.

In 1990 the first IPCC report was presented. These reports began to present historical events of the last decades, providing scientific information with greater quality and reliability. These reports stressed that man is the one who is altering the planet's climate with the exaggerated consumption of natural resources, especially fossil fuels53.

In the 1990s, problems that could compromise the survival of ecosystems became evident. Thus, a great impulse was given with regard to environmental awareness, as countries accepted to pay for the quality of life on the planet. Companies started to worry about more environmentally responsible actions84.

The Conference of the Parties (COP) first met in early 1995 in Berlin. The main objective of the meetings of the Conference of the Parties was to find solutions to the environmental problem of global warming. Since then, several conferences have been held, the most important of which was the COP-3 (Conference of the Parties No. 3), held in December 1997 in Kyoto, Japan, which resulted in the establishment of the Kyoto Protocol53.

One study reports that the developed countries' initial goal of reverting their emissions to 1990 levels by the year 2000, set at the first Conference of the Parties in Berlin, would not be possible. Thus, it was necessary to establish a Resolution called the Berlin Mandate, which aimed to review the pre-established commitments. The Berlin Mandate determined that developed countries should sign quantitative emission reduction targets. In addition, policies and measures were to be created that would be necessary to achieve these targets37.

1.5.1. Main consequences of climate change

Human-induced climate change is responsible for the continuous increase in the global average land temperature, changes in precipitation levels, changes in the frequency and intensity of extreme weather events, and the rise in the mean sea level. According to the IPCC's fifth report, the warming of the Earth's climate is not a misconception, since studies confirm the increase in the Earth's surface and ocean temperatures, the large-scale melting of the polar ice caps, and the global rise in sea level50.

The main consequences of Climate Change are

- Global warming;

- Increased frequency of extreme weather phenomena such as hurricanes, storms and cyclones;

- Rise in the average sea level, which will mean the disappearance of small states and loss of coastal strips. The average sea level has risen in the last 100 years by approximately 15 cm, and studies indicate that the average sea level will continue to rise in the future;

- Loss of ice cover at the poles;

- Changes in the availability of water resources. Although precipitation levels will increase, arid regions will become drier and less water available. In addition, there is a tendency for the availability of drinking water to decrease;

- Changes in ecosystems and loss of biodiversity;

- Desertification with consequent increase of arid regions;

- Impacts on the health and well-being of the human population with an increase in heat-related and insect-borne diseases;

- Forced Migration;

- Interferences in agriculture.

The greatest evidence that we are going through a period of change in our climate is the global increase in the earth's surface temperature. Almost the entire planet experienced an increase in surface temperature between 1901 and 201250.

1.5.2. Vulnerability to climate change

The impact of climate change is felt differently by populations depending on the regions and social categories affected73.

As noted earlier, the poorest and most disadvantaged countries, particularly tropical ones, are most vulnerable to climate change and its consequences7.

1.6. Greenhouse effect, increased emissions of greenhouse gases

The greenhouse effect is a natural phenomenon that exists in nature independently of man's action and economic activities. This effect, is caused by the presence of certain gases in the earth's atmosphere and for this reason these gases are called greenhouse gases. Without the help of the greenhouse effect, the sun's rays would not heat up planet earth enough for it to be habitable, because the planet's average temperature would be around minus 17°C and its surface covered in ice67.

The Greenhouse Effect ensures that the average temperature of the planet currently remains close to the positive 15°C, as without it it would be more or less above 32°C. Furthermore, without the Greenhouse Effect, the planet would be subject to sudden temperature variations between night and day, as happens on the moon and also in deserts74.

The light rays penetrate the atmosphere, reach the surface of the earth and return to space, when they reach the surface of the planet, these rays change their physical characteristics and are transformed into heat. Part of this heat, emitted by the Earth and which is about to return to space, is trapped in the atmosphere precisely because of the presence of greenhouse gases such as: Water vapor ($H20$); Ozone ($O3$); Carbon dioxide ($co2$); Methane ($CH4$); Nitrous oxide ($N20$); Chlorofluorocarbons (CFCs); Hydrofluorocarbons (HFCs); Perfluorocarbons (PFCs) and Sulfur hexafluoride ($SF6$). The greenhouse effect consists of the capture in the atmosphere of part of the heat generated by the interaction of sunlight with the atmosphere and the earth's surface that would otherwise be reflected back into space67.

Plants take carbon dioxide from the air to carry out photosynthesis, when forests are cut down (their wood is converted into products and their vegetation is burned) much of their carbon is released into the atmosphere as carbon dioxide and the destruction of tropical forests alone accounts for approximately 20% of total greenhouse gas emissions from human activities77.

The Kyoto Protocol and the United Nations Framework Convention on Climate Change (UNFCCC) attribute particular importance to forests in the context of climate change because they play an important role in the global carbon balance.

Cutting and burning forests releases into the atmosphere the carbon dioxide contained in the plants, transforming the forest into a source of $co2$ emissions, since deforestation is among the main anthropic actions of carbon emissions. Table 1 shows the contribution of deforestation as a source of GHG emissions89

Table 1. The main anthropogenic sources of greenhouse gases[1]

Gas	Main Anthropogenic Sources	Contributions (%)
CO2		64
CH4	Energy use, deforestation and change Energy production and use, agriculture, landfills, biomass burning and domestic sewage	20
Compounds halogenated	Industry, refrigeration, aerosols, propellants, expanded foams and solvents	10
N2O	Soils with fertilizers, production of acids, biomass burning and fossil fuels	6

1.6.1. Carbon dioxide

Carbon dioxide is the main product of all combustion reactions, and is therefore the largest greenhouse gas emitted into the atmosphere, where it joins the carbon dioxide that already exists naturally. Plants and oceans are the main carbon dioxide (CO_2) sinks, which prevents levels of this pollutant from increasing further. However, since the compensatory mechanisms in this process are not very significant compared to the atmospheric reserve of carbon dioxide, a molecule of it can remain in the atmosphere for more than a century[41].

1.6.2. Methane

It is emitted mainly by the agriculture and waste sectors, a molecule of methane remains in the atmosphere for less than a decade. However, a molecule of this compound absorbs between 20 and 25 times more infrared radiation than a molecule of carbon dioxide.

The concentration of this gas has been increasing since the industrial revolution. However, a reduction in CH4 levels in the atmosphere has been observed since 1990 in Europe due to the application of good environmental practices, especially in the waste sector, regarding the disposal of waste on the ground[20].

1.6.3. Nitrous oxide

The main sources of nitrous oxide are: the industrial sector, soils, agriculture, and oceans. Although this gas is found in low concentrations in the stratosphere, its GWP is high and the lifetime of N2O in the stratosphere is approximately 114 years1.

From 1750 to 2005, N2O concentrations increased by 270 to 319 ppb (parts per billion, a unit of concentration), with the increase over the last two decades being 0.26% per year Much of the N2O emitted into the atmosphere is converted into nitric oxide (NO) and nitrogen dioxide (N2O) which are ozone precursors. Thus, increased emissions of N2O into the atmosphere imply increased ozone concentration48.

1.6.4. Water vapor

The water vapor present in the atmosphere is responsible for two thirds of the natural greenhouse effect (on clear days). Despite the low GWP associated with this gas, the fact that it exists in large quantities in the atmosphere, makes it an important greenhouse gas. With the planet's temperatures increasing due to the greenhouse effect enhanced by anthropogenic GHG emissions, more water from water surfaces tends to evaporate, which increases the concentration of water vapor in the atmosphere. Thus, the Earth's warming process is enhanced98.

1.6.5. Other greenhouse gases

The decrease in the use of CFCs in favor of the preservation of the ozone layer has decreased the concentration of these pollutants in the atmosphere. However, the need to use substitutes for this compound has increased the concentrations of other gases in the atmosphere such as PFCs, HFCs, and SF 6, which are also halogenated compounds and have the particularity of being powerful greenhouse gases with long atmospheric lifetimes. The emissions of these compounds result from the combustion of fossil fuels and evaporation of various hydrocarbons. Ozone is one of the main constituents of the stratosphere and due to its fundamental role in absorbing ultraviolet radiation, it protects the earth's surface from its harmful effects49.

1.6.6. Aerosols

Aerosols are small solid or liquid particles that originate from natural phenomena, such as dust dispersion from storms and volcanic activity, and from anthropogenic processes, such as the burning of fossil fuels. Although they are not GHGs, they are important compounds in the processes that lead to changes in the planet's temperature, since they produce a cooling effect in the atmosphere in certain regions. These compounds are able to reflect sunlight, thus changing the albedo. On the other hand, aerosols and other compounds that originate them (sulfur compounds, for example), by forming condensation nuclei, contribute to the formation of clouds and indirectly to the increase of the albedo (reflection of solar radiation)[45].

1.7 Carbon cycle

The concept of the carbon cycle is a good starting point for the study of sequestration through vegetation. The element carbon is the main constituent of everything organic, and although carbon dioxide (CO_2) represents only 0.35% of the gases that make up the atmosphere, carbon is an element that in recent years has caused profound changes worldwide[33].

The element carbon is found in the atmosphere as a gas originating almost entirely from the breathing process of living beings (79%) by which it completes what is called the "Carbon Cycle" (Figure 1).

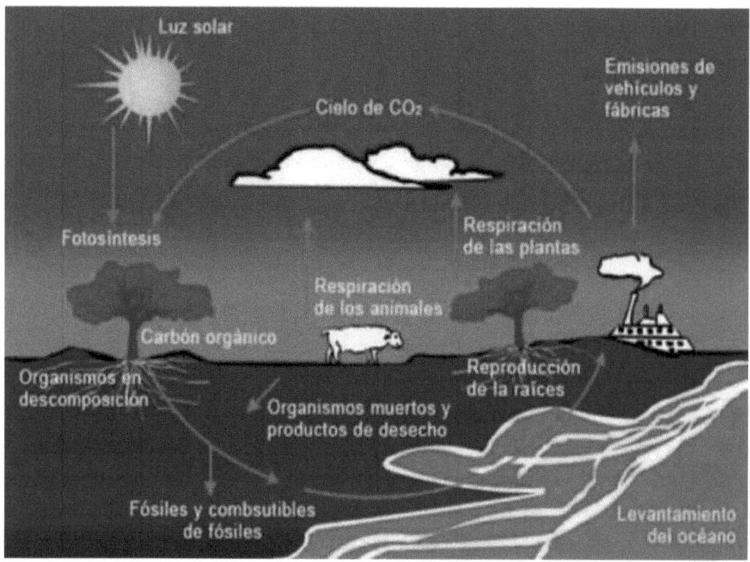

Figure 1. illustration of the carbon cycle

A viable alternative to mitigate the increase in GHGs is the fixation of atmospheric carbon from large-scale reforestation. The four main carbon compartments on earth are: oceans, atmosphere, geological formations containing fossil and mineral carbon, and terrestrial ecosystems[6, 76].

Carbon sequestration is understood to be the photosynthetic capacity that plants have to fix atmospheric CO_2, biosynthesizing it in the form of carbohydrates that are finally deposited in the cell wall[76]. The faster a species grows, the faster it absorbs CO2[4]. Because of the rapid growth of trees in the tropics, one hectare of this forest sequesters much more carbon than one hectare of temperate forest[76].

Among the greenhouse gases that are increasing in concentration, carbon dioxide, methane, and nitrous oxide are the most important. Carbon dioxide stands out with the highest emissions, raising its level of contribution to global warming.

If the CO_2 concentration continues to increase the earth's temperature will result in an increase in sea level and alteration in the variability of hydrological cycles, threatening life on the planet. Forest carbon sequestration is a viable alternative to mitigate the worsening process of global temperature increase by increasing GHG76.

The carbon dynamics in a forest are determined by the assimilation of CO_2 through total photosynthesis; the release of carbon through respiration of autotrophic plants; the transfer of carbon from the soil in the form of leaf litter, wood, and roots; and the eventual release of carbon from the soil back to the atmosphere through decomposition and respiration by microbes and other heterotrophic beings.67

1.7.1. Miombo forests and carbon sequestration

Currently, there are many cases of estimation of the amount of carbon in the aerial part of exotic species88, but the case is different in relation to native forests, in particular those of Miombo.

A carbon sequestration strategy is being developed through agroforestry practices and to reduce emissions derived from deforestation and degradation of Miombo woodlands87. As an example, in some experiments implemented in the carbon sequestration project by the community of Nhambita (Mozambique), farmers have received carbon payments at a rate of 4.5 US dollars per ton of CO_2, which can generate between 433 and 808 US dollars per hectare in seven years28.

Carbon sequestration through land use and forest planting can promote sustainable rural livelihoods and generate reductions in carbon emissions that can be verified by the international community87.

Despite this, little is known about carbon flux and storage in Miombo woodland58, since calculating carbon stocks in dry forests requires a different approach and more research on them compared to wet forests8.

The quality of management to forest plantations is a significant contribution to controlling carbon dioxide levels in the atmosphere. Other activities that can contribute are conservation of forests in danger of deforestation, forest rehabilitation, afforestation, reforestation. These forests have a high capacity to absorb carbon dioxide and therefore to contribute to the reduction of atmospheric carbon dioxide35.

1.7.2. Carbon sequestration by forests

The removal of carbon dioxide from the atmosphere by means of forest plantations is one of the options for offsetting greenhouse gas emissions2. This removal occurs through the process of photosynthesis. Carbon dioxide is fixed in reduced carbon compounds, which are stored in the form of biomass. In turn, through the process of plant respiration, decomposition of its residues and biomass carbonization, the carbon is emitted again and returns to the atmosphere75.

The proportion of carbon absorbed by forests is related to their growth and age. Forests remove carbon in greater proportions when young and growing as they reach maturity and growth stabilizes, uptake is reduced13.

During the initial phase of development of a forest, much of the carbohydrate is directed to the biomass production of the crown and roots. However, as time goes by, the relative biomass production of the trunk increases and that of the leaves and branches gradually decreases, thus forests appear as a great encouragement, because, besides being a renewable natural resource, they can contribute decisively to reduce the environmental impacts of the greenhouse effect and its implications on climate change83.

1.8. Clean Development Mechanism (CDM)

The Clean Development Mechanism (CDM) aims to achieve sustainable development in developing countries through the deployment of cleaner technologies, it is also a mechanism that facilitates compliance with GHG emission reduction targets of developed countries17.

The CDM is a mechanism based on the development of projects that have private initiative in charge. CDM project activities in developing countries must have real, measurable, long-term benefits. In addition, these projects must also be related to reducing GHG emissions or at least influencing the incidence of CO237.

Therefore, CDM projects can involve the substitution of fossil energy by others of renewable origin, rationalization of energy use, forestation and reforestation activities, more efficient urban services, among other possibilities. In addition, these projects must involve one or more of the gases provided for in the Kyoto Protocol.

A CDM project will only be valid if the activities therein will unequivocally contribute to the reduction of GHG emissions, If the emissions will occur anyway, i.e. naturally, this project will not be considered eligible for CDM. This is the complicating point for this type of project, because the confusion comes not from understanding the mechanisms, but in developing the empirical basis57.

The first stage of a CDM project is the PDD (Project Design Document), and its preparation is a crucial point for the continuity of the project. It must be done using a specific form, for large-scale activities. For small scale activities, the form differs in some aspects. Normally this document is prepared by members on their own or with the support of consultants. It is important to remember that the consulting company that supported the elaboration of the PDD cannot participate in the validation phase84.

The PDD is the document that brings together the information that characterizes a project activity. It addresses the technical and organizational aspects of the project activity; it justifies the choice of baseline and monitoring methodology, and it will demonstrate its effectiveness. The PDD must follow the current model established by the UN CDM Executive Board37.

The PDD must include a description of the project activities and participating agents. The information must be clear and transparent: description; technical information and geo-referenced location; baseline methodology adopted; project boundaries; period for obtaining Carbon Credits; methodology adopted for the monitoring plan; calculation of the project's baseline emissions; reference document on the environmental impact assessment of the project; summary of comments from the agents involved in the project's assessment process53.

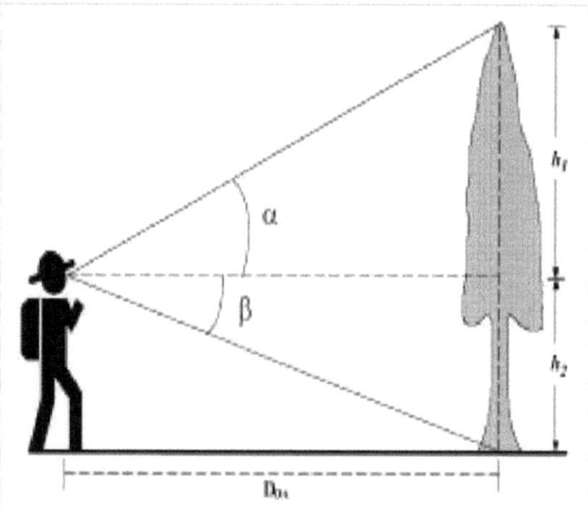

CHAPTER 2. DATA, PROCEDURES AND METHODOLOGY

CHAPTER 2. DATA, PROCEDURES AND METHODOLOGY

2.1. Study area

The study took place between March 2017 and August 2018, in twelve (12) plots located in the Brito Teixeira forest and that of Miombo, both located within the perimeter of the Experimental Station of Chianga (Figure 2), belonging to the Institute of Agronomic Research (IIA), where the Faculty of Agricultural Sciences is located. Chianga is located 13 km NE of the city of Huambo, in the province and municipality with the same name. It has an area of approximately 2,550 ha, defined by the parallels 12° 14'and 12° 16' South latitude and the meridians 15° 48' and 15° 52' East longitude5.

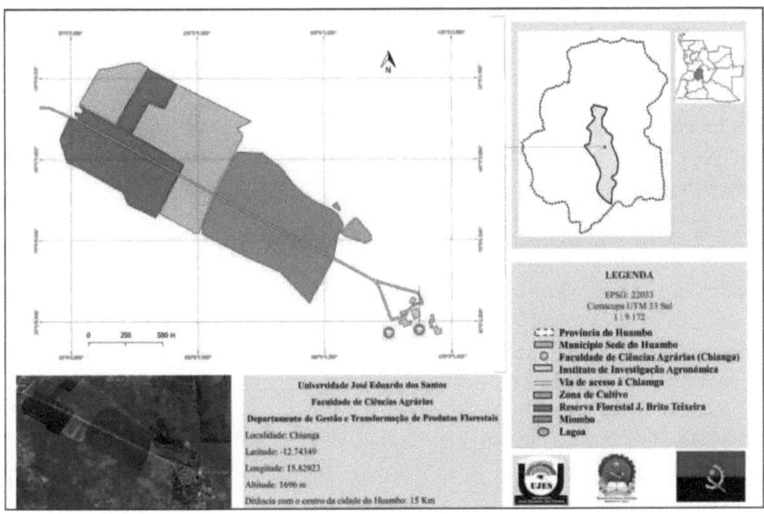

Figure 2. Location of the study area

Chianga is limited to the south and southwest by the Culimaala River, to the northeast by the Caluapanda Creek, to the east by the Chianga Creek, both tributaries of the Culimaala, and to the north and northeast by a broken line that closely follows the Benguela Railway Forest Reserve. The area falls within agro-ecological and socio-economic unit 24, with altitudes above 1500 m. It is characterized by tropical deciduous trees that receive 800 to 1396 mm of precipitation annually. Its average annual temperature is 20°C, and the hottest period is from September to October, which coincides with the beginning of the rainy season with highs of 25°C and 27°C and the coldest month is June with averages of 11°C and 13°C. The average annual relative humidity varies from 60-70% and the minimums in August are 35-70%19.

The Chianga soil is slightly ferralitic yellow or orange in color, derived from eruptitic, crystallophilic, and quartziferous rocks. The soil has a loamy texture and the Ph varies between 5.2 and 5.5. The phosphorus concentration is 43 mgL-1 and the organic matter percentage is 2.2%40.

2.2. *Establishing the sampling plot*

The first step in any study carried out in the field is planning the experimental design. The choice of sampling plot should be based on several aspects, among which are precision, the nature of the information required, and its relative cost56.

A good choice of both the sampling design, the shape of the plots, and the size of the plots is important to ensure representativeness of the entire population in our sampling plots. A Brazilian researcher has broadly classified two groups of sampling methods for field data survey in relation to the type of sample: i) the plot or fixed area method: ii) the quadrat or variable area method56.

The set of methods to be applied in the fixed area can have a single or several plots, but it is recommended to use several smaller plots, rather than a single one with a large sample unit, because it presents among other advantages the advantages of samples with greater precision, the heterogeneity of the vegetation.

The fixed area method with more than one plot was chosen in this work.

Regarding the distribution of the plots, these can be established randomly, systematically or in a combined manner72. In our case, the method chosen was systematic sampling, because it consists in distributing the sampling units in a regular pattern in the study area; this model has the following advantages:

a) Detect variations, plus its application is simpler in the field and allows for better population estimation

b) The selection of sample units, in systematic sampling, is easier and faster;

c) The organization, supervision and checking (remediation) of some sample units becomes operationally easier;

d) The size of the population does not necessarily need to be known;

e) The systematic sample units are more evenly distributed in the population, making it more representative and efficient when there is any trend or concentration of certain characteristics;

f) Systematic sampling is most efficient when you want to map the forest with respect to its typology.

Regarding the shape of the plot, the most commonly used are square and rectangular, due to the ease of installation in the field, although other authors recommend the circular shape, the fact is that currently this is the shape employed for forest inventories by some international organizations26.

It should be noted that, in general, considering the same area for the plots, rectangular shapes are more efficient than isodiametric ones. As the clustering pattern of individuals of the same species tends to be isodiametric, elongated plots are more likely to intersect part of several clusters, while those of equal diameter may coincide over clusters of individuals or over the spaces between them, and a very large number of plots are required to obtain a significant average94.

Based on this background, our choice was rectangular plots, 25x20, of which six were from the Miombo and the other six from the Planted forest (Brito Teixeira)

2.3. Forest Inventory

The inventoried area is characterized by having two types of forest ecosystems, the planted forest consisting of *Eucaliptus grandis* and *Pinus patula* and the natural or Miombo forest consisting mainly of 16 genera and 19 species (see table 1 in the results chapter). Below (Figure 3) can be seen the Brito Teixeira Forest (left) and the Miombo Forest (right) located in Chianga (Huambo).

Figure 3 - Brito Teixeira and Miombo forests

For the inventory activity, 12 rectangular plots were marked (6 in each of the forests) with a size of 25 x 20 m (500 ^{m2}) each, taken systematically (every 5 m) to avoid the edge effect and tree-free zones. While marking these, the geographic coordinates of their centers were extracted.

2.3.1. Estimation of the age of the planted forest and that of miombo wood

The estimated age of the trees *(Eucaliptus grandis* and *Pinus patula)* in the planted forest was based on the studies of a local researcher, who states that the Brito Teixeira forest was established between 1965 and 1968 at the Chianga Agricultural Experiment Station (Huambo, Angola). As for the ages of the trees in Miombo, due to the lack of an auger to estimate the age by the dendrochronology method, the age of the trees was estimated based on the physiology that the trees presented and based on oral sources, due to the fact that this forest was put under a silvicultural measure based on the use of fire between the years 1980 and 199055.

2.3.2. Measuring and recording data

All trees that had a normal diameter (ND) greater than or equal to 4 cm were marked in each plot.

2.3.3. Determination of dasometric variables

Prior to measuring the dasometric variables (density, diameter of normal section, height, basimetric area, biomass and volume), we marked the central coordinates in each of the twelve (12) plots with a GPS *(Garmin etrex 20x GPS)*.

2.3.4. Normal section diameter

The diameter within each sub-plot, in a total of 4 per plot, was determined on the normal section at 1.30 m height (dn) of all trees inventoried (dn > 4 cm) using a Haglof "Mantax Black" fork with nailable jaws (l=650 mm); for those trees with very large diameters that could not be measured with the fork, a tape measure was used.

2.3.5. Density

Density, defined as the number of individuals per unit area, was determined for all species found in the three sampling plots. The objective was to identify the number of the main species present in the study areas as shown in Appendix III and IV.

2.3.6. Height of Trees (Ht)

In the twelve (12) study plots for trees with Dn > 4cm at 1.30 m height, height was measured with a Clinometer that allowed for accurate height measurement, with the aid of a sighting scope at a fixed distance from the tree to the observer equal to or greater than the height of the tree. In the trees that it was possible due to the low height of the individuals, the measurement was made with the help of a tape measure.

2.3.7. Identification of species

In each of the twelve plots most of the species found were identified, according to table 2.3, by their local name in the Umbundo language and by their scientific names. For the botanical identification we used works carried out by researchers in the region[19], [82].

2.3.8. Volume determination

The volumes were estimated, considering the basimetric area at a height of 1.3 m, by the formula Wood volume (^{m3}) = n / 4 x Diameter2 x Height x Fe. In this case Fe is an empirical form factor that in the Miombo species should be between 0.36 for heights above 12 meters and 0.39 for heights below 12 meters and in the pine and Eucalyptus species is 0.33.

2.3.9. Shell thickness determination

To determine the thickness of the bark, a hole was made in the trunk of the tree with the help of a knife and then the measurement was made with a 30 cm graduated ruler.

2.3.10. Determination of branch length (cm)

To determine the branches, the three largest branches of each tree were selected and the height of each branch was measured with the help of a tape measure, and then the values were added, and an average of the height of the branches of each tree was finally found and used.

2.3.11. Determination of crown diameter (cm)

We measured the branches with a tape measure from the south end to the north end and from the east end to the west end, then added the results and divided by two, we worked with the average of each tree.

2.3.12. Creation of the database in Excel and Acquisition of the Canadian mathematical model to estimate carbon sequestration per tree.

After conducting the forest inventory, the data collected in the field was entered into an Excel spreadsheet (creation of the dasometrics database) and the treatment of these data. It was also necessary to acquire the Canadian mathematical model *The Envirothon* and *The Yukon Envirothon* to estimate carbon sequestration by trees.

Thus, the inventory and analyzed geometric data were introduced in the model to estimate the net biomass rate per tree, annual biomass rate and the amount of carbon sequestered per tree. The grouping of these individual rates by the values of trees from the same plot and species allowed us to know the average carbon sequestration rate per plot and per species.

In summary, the methodological steps followed in this work are presented in Figure 4.

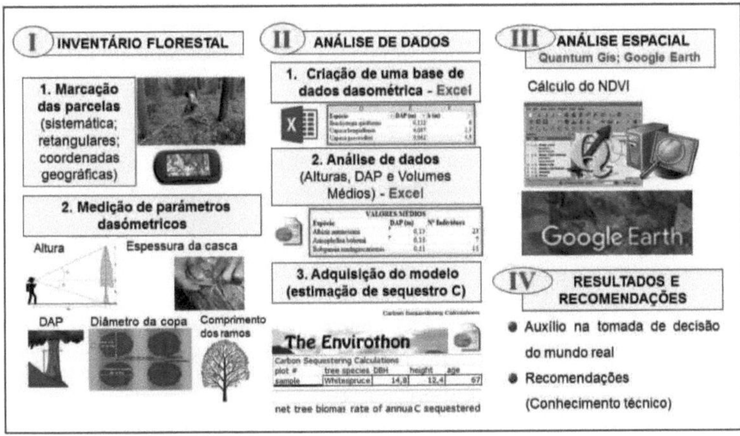

Thus, results were obtained (described in the next chapter), which led to the conclusions and recommendations described in the respective chapters.

2.4. Calculation of the Normalized Difference Vegetation Index (NDVI)

Meanwhile, based on the remote sensing data, the Normalized Difference Vegetation Index (NDVI) was calculated, which according to studies conducted in a local context, is a useful indicator in analyzing the presence or absence of vegetation or biomass and photosynthetic activity69.

2.5. Ways to estimate carbon sequestration and the main formulas

Some authors point to three ways of quantifying the carbon stored by natural ecosystems: spectral data, data from field and statistical models99. The methodology used in this work consisted of the use of field data (inventory) and statistical models based on the model proposed by the *Canadian Forest Services,* to evaluate the potential carbon stock sequestered by the Brito Teixeira and Miombo (Chianga, Huambo) forests. The main formulas of the Canadian Forest *Services* model are12:

Canopy **Cover Fraction**
(%) =
Unit diameter of copax
100 / Total diameter of copax

Net biomass rate
(t) = (Aboveground biomass x0
.26) + Aboveground

Wood volume
(m3) = n / 4 x Diameter2 x Height x f

Annual
biomass **rate**
(t/year) =Net biomass rate/
Tree age

Above ground biomass (t) =
Bark thickness +Branch length
+ FCD+
Volume

Amount of carbon
sequestered
(t) = Annual biomass rate
/ 2

Two methods of estimating forest biomass can be adopted. The first is the direct method, which involves a direct count by cutting down a sample of trees and weighing the biomass components. Another method is the indirect method, more used today, and consists of estimates through allometric equations that include variables such as diameter and height of trees, whose values are obtained through forest inventories, and has the advantage of keeping the tree alive86.

In addition to volume and height values, it is useful to have biomass data. The biomass value is used with a conversion factor of 0.5 to estimate the value of stored carbon, since carbon mass represents approximately 50% of biomass3. The conversion factor most commonly used in studies and in this one in particular is 0.554.

CHAPTER 3. CARBON SEQUESTRATION BY PLANTED AND NATIVE FORESTS IN HUAMBO

CHAPTER 3. CARBON SEQUESTRATION BY PLANTED AND NATIVE FORESTS IN HUAMBO

3.1. Forest Inventory

The species with the highest number of individuals (Table 2) was *Pinus patula* with about 20 %, followed by *Eucalyptus grandis (17 %), Brachystegia spiciformis (14 %)* and *Uapaca benguellensis* (8 %). Also in the same table can be seen the predominant Miombo species in the study area, which shows a great diversity of species, which favors a rich biodiversity. This is in accordance with other authors36, who consider the Angolan Miombo very diverse, with the following main species: *Julbernardia paniculata, Brachystegia bohemii, Bobgunnia madagascariensis, Albizia adianthifolia, Pericopsis angolensis, Syzygium guineense, Pterocarpus angolensis, Anysophyllea bohemii* and *Isoberlinia angolensis.*

The plot with the largest number of individuals is Parcel 6 of Miombo with 51 trees, followed by Parcels 1 and 5 of Miombo (49 individuals each) and Parcel 2 of Miombo (41 trees); the plots with the fewest individuals were those of the Brito Teixeira Forest, Parcels 6 and 5 with 15 and 22 individuals, namely as shown in Table 2.

As for DBH as can be seen in Table 2, the species with the highest value of this variable are: *P. patula* and *E. grandis* with approximately 0.3 m, followed by *H. acida* (0.21 m) and *I. angolensis* and *B. spiciformis* with 0.17 m each. At the plot level, Plot 5 of the Brito Teixeira Forest has the largest diameter (0.35 m), followed by Plots 6 (0.32 m) and 2 (0.30 m) of the same forest (Figure 5d).

The species with the greatest height (Table 2) was *Pinus patula with an* average of 21 m, followed by *Eucalyptus grandis* with an average of 20 m, *Brachystegia spiciformis (9.3 m), Isoberlinia angolensis* (6.1 m) and *Syzygium guineense* (5.5 m). The species with the lowest height were *Terminalia brachystemma* (1.7 m), followed by *Psorospermum febrifugum (3 m) and Ochna schweinfurthiana* (3.2 m).

The plots in the Brito Teixeira forest, showed higher average height, among them the following stood out: Plot 2 (21.6 m), followed by Plots 6 (21.4 m) and 1 (21 m), as shown in Figure 5e. Lower average height values were observed in the Miombo plots (Figure 5e), which is associated with the physiognomy of the species of this forest.

Table 2. Average values of DBH, Height and Number of Individuals per species

Species	N° Individuals	DBH (m)	h(m)	Volume (m^3)	Age (years)	Peel thickness (cm)	Branches (cm)	D canopy (cm)
Albizia antunesiana	23	0,12	4,3	0,016	15,6	0,478	253	363
Anisophyllea bohemii	7	0,13	5,1	0,022	12,6	0,392	162,1	271,4
Bobgunnia madagascariensi	11	0,11	4,5	0,014	14,5	0,418	200	290
Brachystegia bohemii	17	0,12	4,8,	0,018	15,3	0,566	245,7	544,7
Brachystegia longuiflora	18	0,1	4,7	0,012	12,8	0,461	212,6	324,4
Brachystegia spiciformis	57	0,17	9,3	0,069	18,1	0,73	291,8	513
Hymenocardia acida	7	0,21	3,8	0,044	12,6	0,385	234,2	352,8
Isoberlinia angolensis	13	0,17	6,1	0,046	16,5	0,547	273,5	393,5
Monote spp	8	0,1	3,5	0,009	15,3	0,512	209,1	436
Ochna schweinfurthiana	9	0,06	3,2	0,003	13	0,333	172,8	273,3
Parinari curatelifolia	12	0,1	5,4	0,014	13,4	0,427	183,3	295
Pericopsis angolensis	1	0,056	3,4	0,003	12	0,3	120	261
Psorospermum febrifugum	2	0,1	3,0	0,008	12,5	0,55	165	210
Rothmannia engleriana	2	0,1	3,8	0,010	14	0,45	235	308
Sterculia quinquebola	3	0,1	4,5	0,012	12,6	0,366	132	13,3
Syzygium guineense	13	0,13	5,5	0,024	14	0,383	185,4	255,3
Terminalia brachystemma	6	0,08	1,7	0,003	11	0,3	153,8	266,6
Uapaca benguellensis	32	0,1	3,6	0,009	13	0,4	151,6	247,6
Uapaca gosswoilleri	9	0,11	3,8	0,012	12	0,6	159,4	278,8
Eucalyptus grandis	67	0,3	20,3	0,474	48,5	1,98	372,6	565,4
Pinus patula	79	0,3	21,1	0,492	52	2	349,7	561,7

As for age as shown in Table 1, the species with the highest value are: *P. patula (52 years), E. grandis with (48.5 years), Brachystegia spiciformis (18 years), and Isoberlinia angolensis (16.5 years)*. As for age, the following Plots 1,2, 4, and 6 of the Planted Forest stood out with the greatest age, with an average of 52 years per plot. The plots that presented the lowest age were Plots 2 and 3 of the Miombo, with an average of 14 years for each one (Figure 5c). The ages of the trees correspond to those found in the study of other local researchers55 on forest covers and the recovery of fertility of ferralitic soils in Angola.

Regarding bark thickness, the most prominent species were: *Pinus patula (2 cm), Eucalyptus grandis (1.98 cm), Brachystegia spiciformis (0.73 cm), Uapaca gosswoilleri (0.6 cm) and Brachystegia bohemii (0.56 cm)*. It is observed that Parcel 1 of Brito Teixeira with 2.3 cm presents the highest average value of bark thickness, followed by Parcel 4 of Brito Teixeira (2.2 cm). Parcel 4 of Miombo (0.28 cm) had the lowest average bark thickness.

The species with the longest branch lengths are described below: *Eucalyptus grandis with 372.6 cm, followed by Pinus patula 349.7cm, Brachystegia spiciformis* 291.8 cm, *Isoberlinia angolensis* 273.5 cm. Plot 5 of Brito Teixeira Forest showed the highest average branch length value (485 cm), followed by Plots 1 (465 cm), and 2 (430cm) of the same forest. The Miombo plots, on the other hand, showed lower average branch length values.

It should be noted that the species that presented the highest value of average crown diameter were the following: *Eucalyptus grandis with* 565.4 cm, *Pinus patula* (561.7 cm), *Brachystegia bohemii* (547.7 cm) *and Brachystegia spiciformis* (513 cm). According to Figure 5b, the plots with the highest value of mean canopy diameter are: Plots 5 (700 cm), 1 (685 cm) and 2 (660 cm) from Brito Teixeira Forest. The plots with the smallest canopy diameter were from Miombo (Figure 5b).

The evaluation of the dasometric data showed that there is a significant difference between the data from the exotic forest and the native forest; this difference has already been observed locally by different authors55, 69.

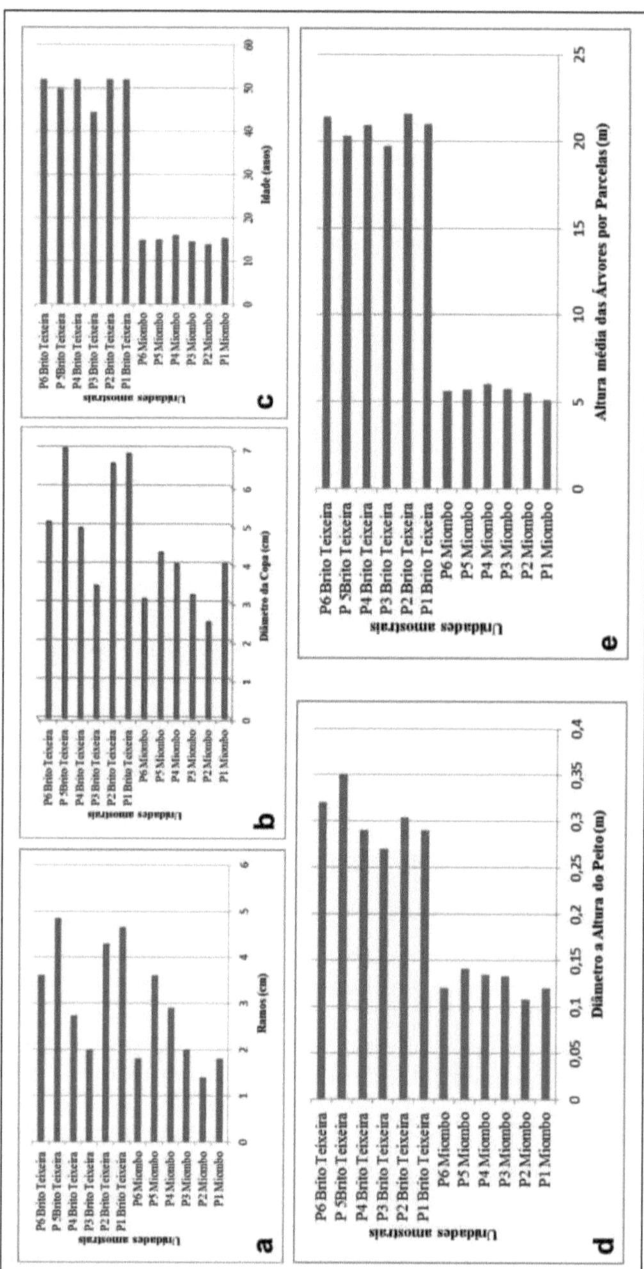

Figure 5: Mean values of dasometric data per plot

3.2. Spatial distribution of biomass and carbon

The Normalized Difference Vegetation Index (NDVI) obtained showed that there was greater photosynthetic activity in Miombo zones relative to the perimeter of the planted forest. In Figure 6, values closer to 0 (zero) indicate the absence of vegetation, little photosynthetic activity or exposed soil, while values close to 1 indicate a large amount of photosynthetically active vegetation. These values resemble those presented by local researchers69.

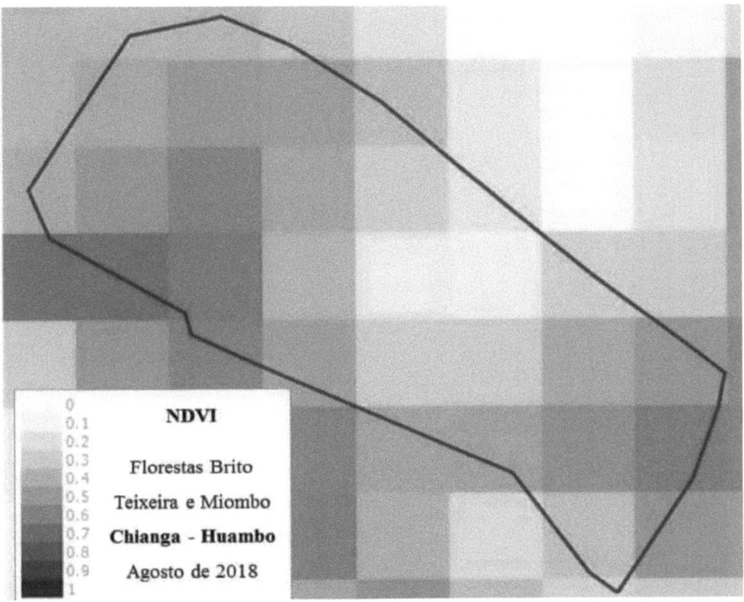

Figure 6. NDVI and phenological status of Brito Teixeira and Miombo (Chianga) Forests

In particular, the areas with the most photosynthetically active vegetation correspond to the areas with the highest carbon sequestration rates.

3.3. Canopy Cover Fraction

The canopy cover fraction for *Pinus patula and Eucalyptus grandis* species showed quite significant differences in percentage terms compared to the Miombo species (Figure 7), this is due to the longer length of the branches of these exotic species, which is associated with their physiognomy. A study on Miombo forest typification in Huambo69 showed low canopy cover fraction values for Miombo species.

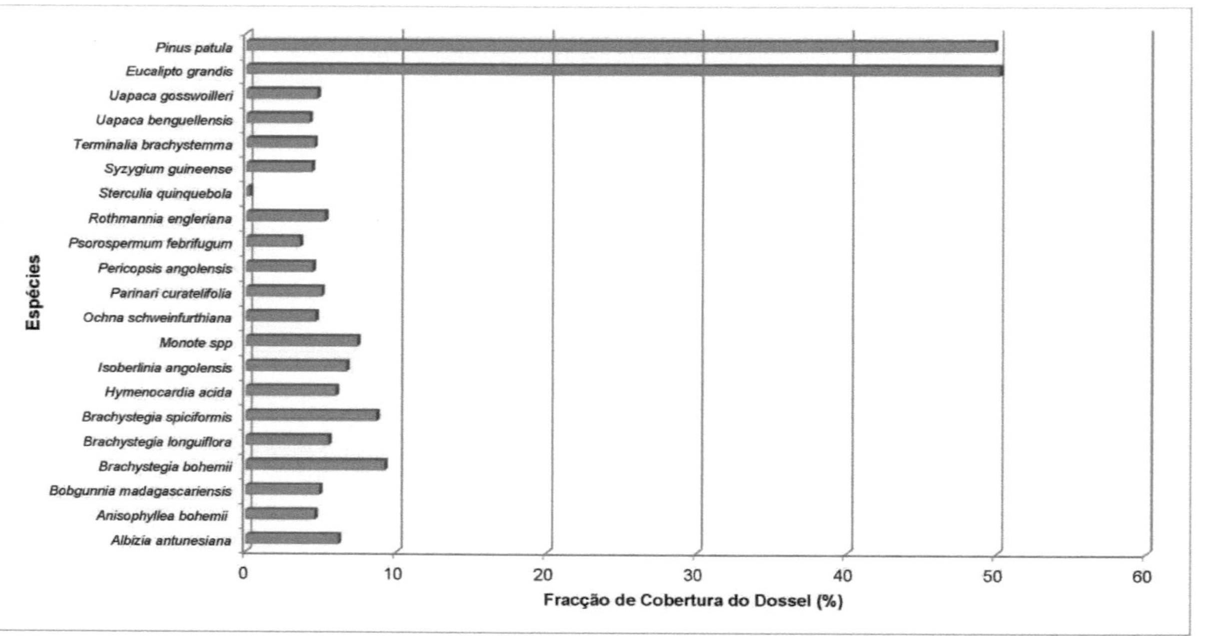

Figure 7. Canopy cover fraction by species

48

3.4. Aerial Biomass

Observing the graph below (Figure 8), one can easily see that in terms of aerial biomass, the *Eucalyptus grandis* presents the highest value, followed by *Pinus patula, Brachystegia spiciformis, Isoberlinia angolensis* and *Albizia antuneziana,* these values can be justified by the fact that these species presented the greatest diameter and the biomass is directly linked to the diameter of the tree. In a study on wood production analysis in Mozambique, it was observed that exotic species presented higher aerial biomass values than native species34.

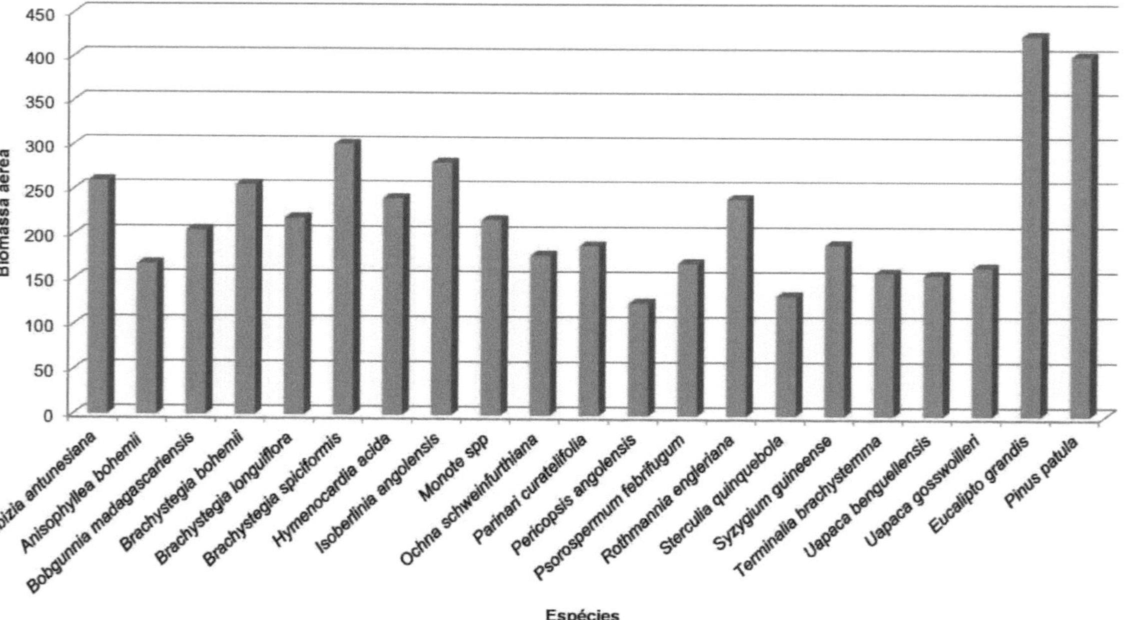

Figure 8. Above ground biomass by species

3.5. Net biomass rate

As for the net biomass, it can be seen in figure 9 that the *Eucalyptus grandis* and *Pinus patula,* showed high values compared to the Miombo species, this can be justified by the fact that the exotic species in this study have shown larger diameters compared to the Miombo species, as well as the age of the two forests, since some species develop more the older they reach.

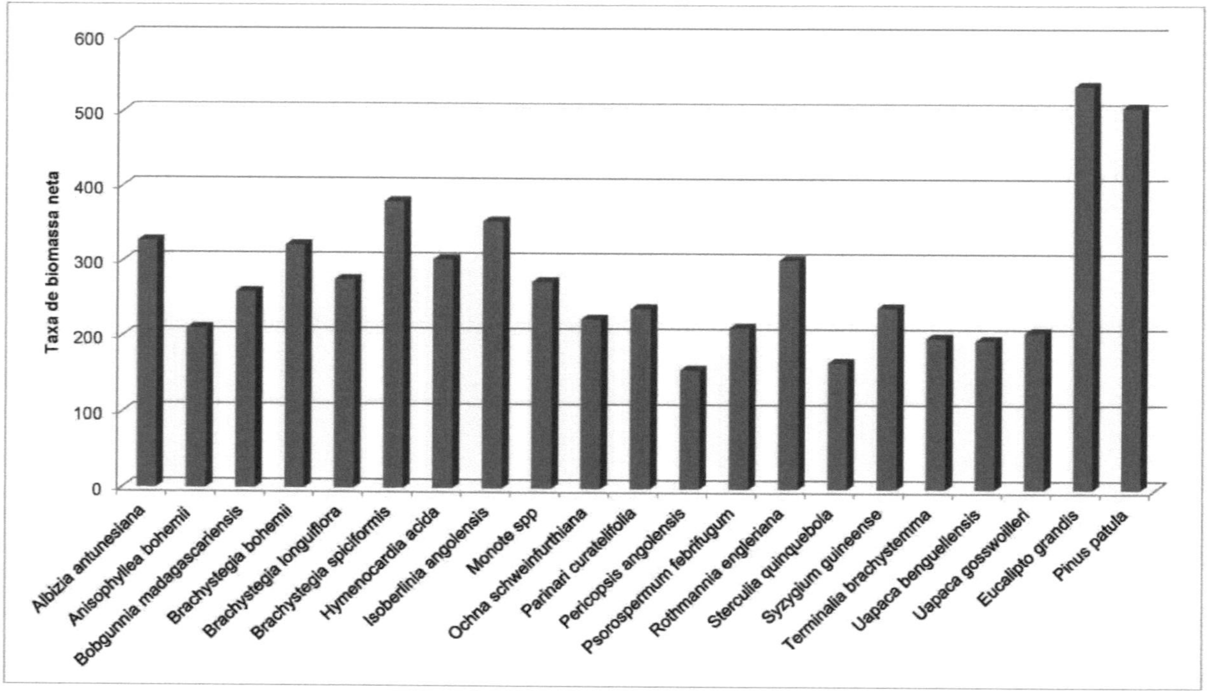

Figure 9. Net biomass rate by species

52

Results on higher biomass rates as a function of age were observed by a study in a tropical context85, which obtained total biomass production that ranged from 3.33 Mg.ha-1 at 12 months to 75.35 Mg.ha-1 at 48 months of age.

3.6. Annual Biomass Rate

Regarding the annual rate of biomass, it can be observed that the miombo species, with particular emphasis on *Hymenocardia acida, Rothmania engleriana* and *Isoberlinia angolensis, showed the* highest values compared to other species including *Eucalyptus grandis* and *Pinus Patula,* which had a lower annual rate of biomass (Figure 10), these results can be justified by the fact that the species of miombo in this study had lower ages compared with the species of *Eucalyptus grandis* and *Pinus patula,* which had higher ages.

This result is in agreement with a study4, which points out that the faster the growth, the faster the absorption of carbon dioxide, and because the exotic species in our study are older, in this case the photosynthetic function of the same is greatly reduced compared to the miombo species that were younger. If the study was done with species of the same age, there we would have the exotic species with the highest annual rate of Biomass.

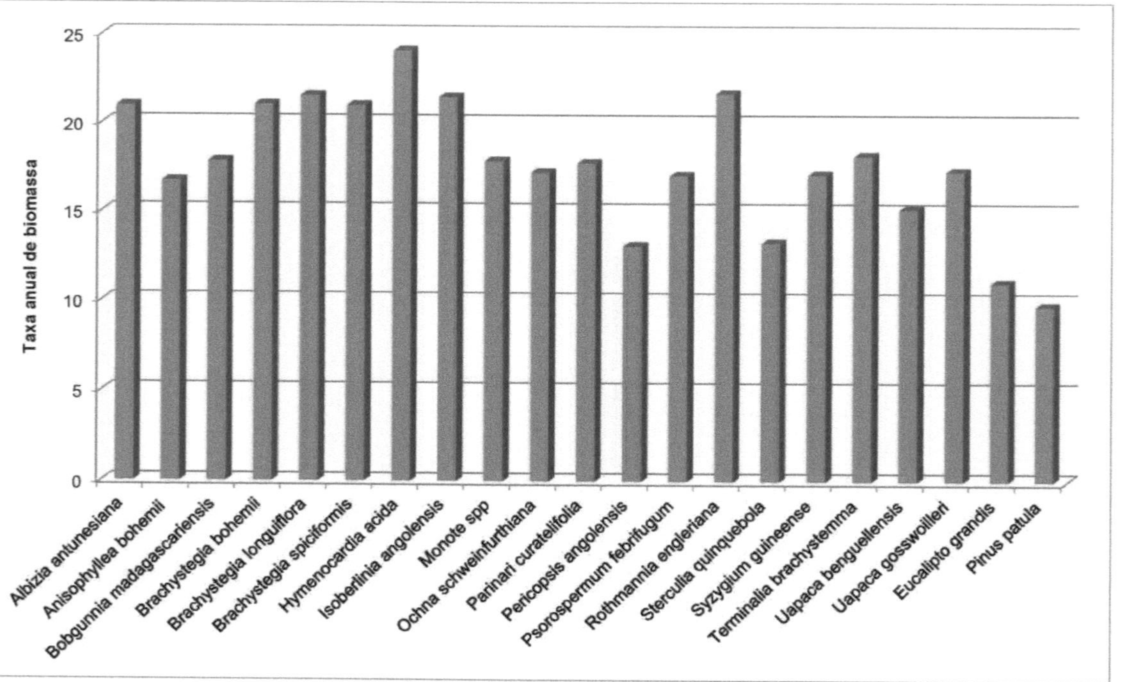

Figure 10. Annual rate of biomass produced by species

54

3.7. Carbon Sequestration

In this study as can be seen in Figure 11 and Appendix II, the Miombo forest showed higher carbon sequestration rates compared to the planted forest. The basis for this was considered to be the average age of the Miombo trees (15 years) compared to the trees in the Brito Teixeira Forest (50 years). Thus, among the species that showed the highest capacity for carbon sequestration throughout the year are: *Pericopsis angolensis* (1679 Mg(C) ha-1, which represents 16 %), *Ochna schweinfurthiana* (533 Mg(C) ha-1, 12 %), *Terminalia brachystemma* (274 Mg(C) ha-1, 6.3 %) and *Parinari curatelifolia* (257 Mg(C) ha-1, 6 %).

Some researchers performed statistical analyses that showed the existence of differences between the values of carbon sequestered by two different forests. The authors found that the foliage was the plant part that showed the highest value of carbon contents. However, as in the present study, the carbon contents were lower than those suggested by the IPCC (*Intergovernmental Panel on Climate Change*)[97].

Already some researchers consider101 that in conserved forest areas, the above ground living biomass has the largest share in the total carbon stock, mainly due to the large trees.

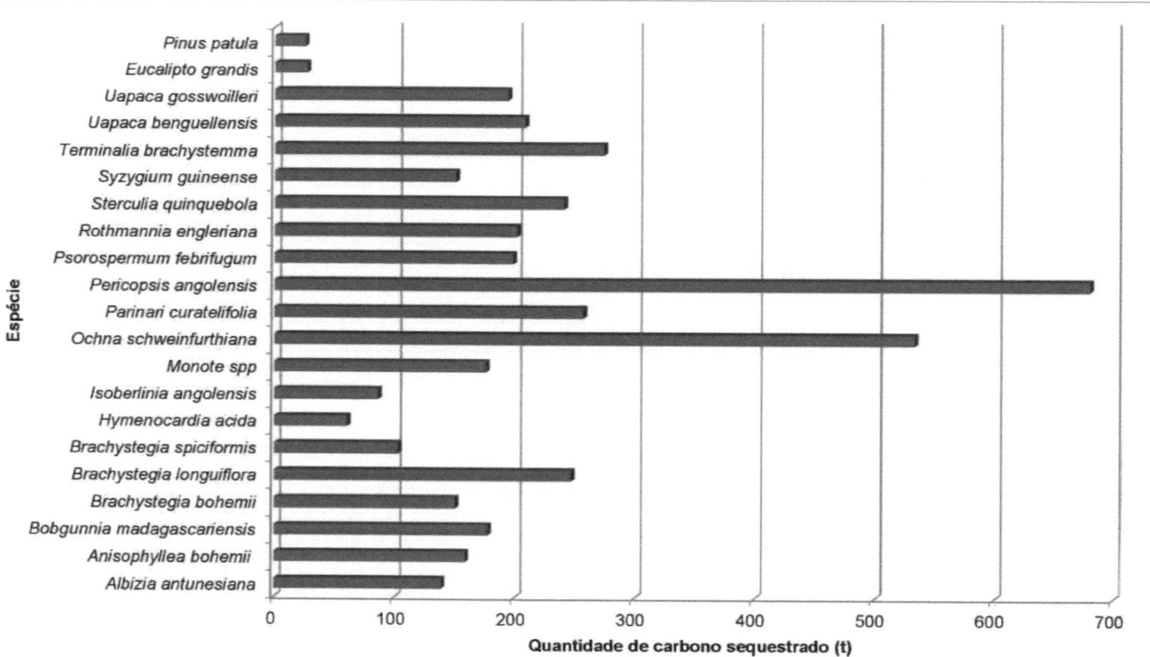

Figure 11. Sequestered carbon by species

Different results from this study were observed in Mediterranean92 areas, which considers Portuguese forests to have a carbon sequestration capacity that, at the ecosystem or lifetime level, can reach high amounts. Although at the country level, sequestration can reduce if the level of harvest, fires and other activities that decrease the *stock,* exceed the sequestration capacity without compensation (for example with new plantations or change in management options).

Although the World Health Organization considers that in Angola of the total forest area, only 24 000 km2 are classified as forests of high biological diversity and greater capacity for carbon sequestration, mostly located in the provinces of Cabinda, Zaire, Bengo, Cuanza-Norte and Uige, the truth is that this idea can be contrasted by the fact that the younger forests in the central and southern regions of the country have greater capacity for carbon sequestration. However, further studies should be conducted in order to ascertain the facts100.

These last parameters (Canopy Cover Fraction, Aerial Biomass, Wood Volume, Net Biomass Rate, Annual Biomass Rate, and Carbon Sequestration) were analyzed at the plot level (as described below), which showed the same significant difference between exotic and natural forest, a reality also observed by other scholars55, [69].

Figure 12a and Annex II show that with regard to the volume of wood, the species on the planted plots were the ones that showed significant differences compared to the Miombo species. This difference can be justified by the fact that the plots containing the planted species had larger diameters and heights when compared to the Miombo species in the study, which had smaller diameters and heights.

The plots with planted species (Brito Teixeira) were those with the highest canopy cover values compared to the Miombo species (Figure 12b and Appendix II) and this may be associated with the longer branches that exotic species have compared to Miombo species. Some researchers consider that one of the different forest compartments that plays an important role in carbon stock is canopy openness, due to the fact that they are directly proportional101.

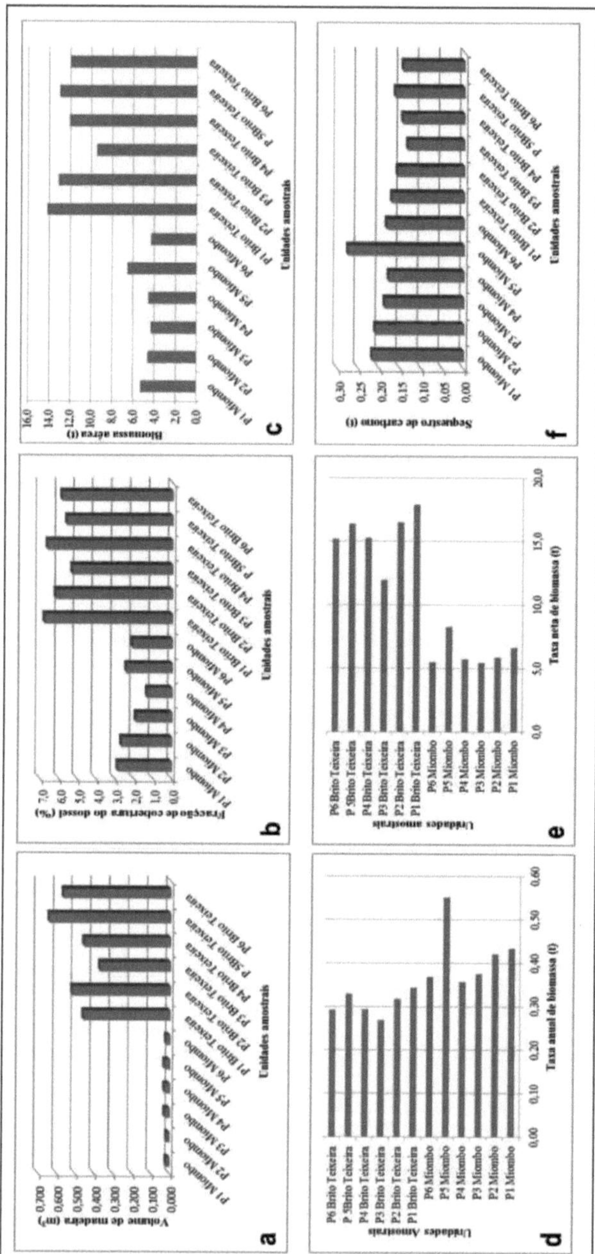

Figure 12. Mean values of geometry and biomass per plot

Higher aerial biomass values were observed in the Brito Teixeira Forest plots, compared to the Miombo plots (Figure 12c and Appendix II). This may be due to the fact that the plots of the planted species show greater diameter and height, whereas the Miombo species for this study were the ones that showed smaller diameter and lower heights compared to the planted species. Similarly, the graph in Figure 12e and the values in Appendix II, show that there is a significant difference in net biomass rate between the plots of the planted forest (with higher values) and the Miombo (with lower values). As for the annual biomass rate, which was different, the Miombo plots showed higher values (Figure 12d and Appendix II).

The Miombo plots sequestered the most carbon and the plots with planted species sequestered the least carbon (Figure 12f and Appendix II). This can be argued by the fact that the Miombo plots in this study had an average age of less than 20 years and the plots with planted species had an average age of over 40 years. In other words, the plots of the Brito Teixeira Forest, being more mature than those of Miombo, sequestered less carbon than the latter. This is in agreement with the IPCC46, which has a methodology for calculating carbon sequestration that does not consider the linear growth of the tree, because in the first years of life its growth is faster and, theoretically, the tree captures more carbon. Thus, the calculation is divided by period, for vegetated areas up to 20 years old and then for areas older than 20 years, which makes the results more accurate.

This official methodology estimates that in its first 20 years of life a hectare of rainforest can capture almost 26 tons of CO_2 per year and after this period, 7.3 tons of CO2 per year. Therefore, a tree can capture about 15.6 kilograms of CO_2 per year in the first 20 years and 4.4 kilograms after that period. Estimating that the tree has a life span of 40 years, it will be able to sequester 667 kg of carbon during its life.

It is important to note that in addition to attempts to contain CO_2 emissions from various industries using fossil fuels and other resources, several processes to reduce global warming are being proposed, especially the conservation of native forests

FINAL IDEAS

FINAL IDEAS

The main conclusions of this study are:

- *P. patula* (20 %), *E. grandis* (17 %) and *B. spiciformis* (14 %) were the species with the highest abundance index;

- The exotic forest showed higher wood volume (average of 0.503 m^3) and biomass (average of 16 t) compared to Miombo (0.023 m^3 and 6 t);

- The trees in the exotic forest (50 years old) are older than those in the Miombo (15 years old);

- Higher rates of carbon sequestration were observed in the younger forest (Miombo), than in the older one;

- The species with the highest carbon sequestration capacity throughout the year are: *P. angolensis* (1679 Mg(C) ha-1, 16 %), *O. schweinfurthiana* (533 Mg(C) ha-1, 12 %), *T. brachystemma* (274 Mg(C) ha-1, 6.3 %) and *P. curatelifolia* (257 Mg(C) ha-1, 6 %).

It is recommended that:

- There should be more conservation of the Miombo, due to its environmental role;

- Establish small conservation areas in areas of the Miombo, in order to guarantee the sustainability of the environmental function of this forest;

- That policies be created so that exotic forests are valued for their environmental role and not just their commercial role.

BIBLIOGRAPHIC REFERENCES

BIBLIOGRAPHIC REFERENCES

1. AEA. (1998). Europe's Environment, Report on Changing

on the State of the Pan-European Environment, prepared as a follow-up to The
Environment in Europe, Office for Official Publications of the European
Communities, Denmark.

2. Albrecht, A. & Kandji, S.T. (2003) Carbon sequestration in tropical

agroforestry systems. Agriculture, Ecosystems and Environment, Amsterdam, v. 99,
(1), p. 15-27.

3. Alves, A. M.; Pereira, J. S. & Correia, A. V.. (2012). Sivicultura, A

Management of Forest Ecosystems. Calouste Gulbenkian Foundation. Avenida de
Berna, Lisbon. 597 pp. ISBN: 978- 97231-1460-7.

4. Baird, C. (2002). Environmental chemistry. Understanding climate change: a
beginner's guide to the United Nations Framework Convention and its Kyoto
Protocol. (2aed). Porto Alegre: Bookman. Brazil.

5. Barbosa, L. A., Grandvaux. (2009). Carta Fitogeográfica de Angola. ASSESCA-
PLP.Coimbra.

6. Barreto, L. V., Freitas, A. C. S. & Paiva, L. C. (2009). Carbon sequestration.
Universidade Estadual do Sudoeste da Bahia. Brazil. 10p.

7. Black, R., Adger, W., Arnell, N., Dercon, S., Geddes, A & Thomas,

D. (2011). "The effect of environmental change on human migration". Global
Environmental Change, 21, p.3-11

8. Blackie, R., et al. (2014). Tropical dry Forests: the State of Global

Knowledge and Recommendations for Future Research. Bogor, Indonesia: Center for
International Forestry Research.

9. Burrough, S. L., Thomas, D. S. G., Orijemie, E. A., & Willis, K. J. (2015).
Landscape sensitivity & ecological change in western Zambia: The long-term
perspective from Dambo cut-and-fill sediments. Journal of Quaternary Science,
30 (1), p 44-58. ISSN 0267- 8179.

10. C.C.E. (2010). Green Paper on Forest Protection and Information in the EU:
Preparing forests for climate change, Brussels.

11. Celender, N. (1983). Miombo Woodland in Africa- Distribution, Ecology and

PatternVersitet. International Rural Development Center. Arbestsapport, Workingpaper 16. Uppsala Suecia. 54 p.

12. CFS - Canadian Forest Services. (2015). Carbon Sequestering

Calculations. Natural Resources Canada. Forestry: Canada. Date accessed: 10/08/2018. Available at:

< https://www.yukonenvirothon.com/carbon-sequestering- calculations.html>.

13. Ciesla, W.M. (995). Climate change and forest management: an overview. Rome: FAO, 128 p. (Forestry Paper; 126).

14. Coelho, A. R. G., et al. (2008). The Commercialization and

Accounting for Carbon Credits Based on Clean Development Mechanism Projects. Access Date:09/14/2018.   Disponívelem   :

< http://www.atena.org.br/revista/ojs-2.2.3-06/index.php/pensarcontabil/article/viewFile/97/97>.

15. Cuesta, A. (2011). Notebooks: the forests of Africa.

subsahariana. Cuaderno Bosques, Marzo - april 2011, Vol 1. XXV, n° (2). ISSN: 1136-0984. Available at:

< http://www.africafundacion.org>.

16. DDA. (2013). Dicionário Digital Aurélio, V.05.

17. Delgado, M. F., & Altheman, E. (2007). Study on the financial viability of the Carbon market. Unopar Cient., Ciênc. Juríd. Empres., Londrina, v. 8, p. 39-48, mar. Available at: < http://bdjur.stj.jus.br/xmlui/bitstream/handle/2011/36695/estudo _about_v

iabilidade_delgado.p df?sequence=1 >. Access date: 10/15/2017.

18. Dieterle, G. (2010), "Sustaining the World's Forests: Managing Competing Demands for a Vital Resource: The Role of the World Bank" in P. Spathelf (ed.), Sustainable Forest Management in a Changing World: A European Perspective, Managing Forest Ecosystems 19, Springer Dordrecht Heidelberg London New York, pp. 9-32.

19. Diniz, A. C. (2006). Características Mesológicas de Angola. (2a), Lisbon: Instituto Português de Apoyo al Desarrolo. ISBN 9728975-02-3.

20. EEA - European Environment Agency. (2009). Annual European Community greenhouse gás inventory 1990-2007 and inventory report 2009 - Submission to the 21.UNFCCC Secretariat. Technical report No 04/2009. (Disponível em:

http://www.eea.europa.eu).

22. ENPRF. (2011). Estratégia Nacional de Povoamento e Repovoamento Florestal. Project for Spatial Assessment of the State of the Forests and Sensitive Ecosystems of Angola. Forestry Development Institute.

23. FAO. (2001a). Global Forest Resources Assessment 2000. FAO Forestry Paper 140. Rome, Food and Agriculture Organization. Disponível em: http://www.fao.org/forestry/ fo/fra/ [Geo-2-394]

24. FAO. (2005). The Global Forest resources Assessment 2005 (FRA 2005). Rome: Forest Resources Assessment Programme. Working Paper 83/S Global forest resources assessment update to 2005. http//www.fao.org/forestry/9690-1-0.pdf.

25. FAO. (2009b), Monitoring and Assessment of National Forest Resources in Angola - Guide for data collection. National Forest Monitoring and Assessment Working Paper NFMA XX/P. Rome, Luanda (Angola).

26. FAO. (2009). Natural Resources. Status of the world's forests, ftp://ftp.fao.org/docrep/fao/011/i0765/ i0765s14.pdf.

27. FAO. (2009). Deforestation: indicators of pressure situation

response.   Available at   :

http://www.fao.org/ag/againfo/programmes/pt/lead/toolbox/Grazi ng/DeforeEA.htm Access Date: 8/09/ 2018

28. FAO - Food and Agriculture Organization of the United Nations. (2010). "Climate-smart" agriculture: policies, practices and financing for food security, adaptation and mitigation. Available at: < www.wocat.org>.

29. FAO. (2010). Food And Agriculture Organization Of The United

Nations 2010. Global Forest Resources Assessment (2010). Rome, Italy. Available at   : < http://www.fao.org/forestry/fra/fra2010/en/>. Date of Access: 14/09/2018

30. FAO. (2012). Rome. FRA 2015. Terms And Definitions. Forest Resources Assessment Working Paper 180. Disponível em: www.fao.org.

31. FAO. (2015). Food and Agriculture Organization of the United Nations. Global Forest Resources Assessment 2015 ^How Are the World's Forests Changing? Rome, 2015. p 4 e 17. ISBN 978-92-5308821-8. Available at: http://www.bivica.org/upload/recursos- forest-evaluacion.pdf. Date de Acesso: 13/07/2018

32. FAO. (2016). The State of the World's Forests 2016. Forests and agriculture:

challenges and opportunities in relation to land use. Rome. ISBN 978-92-5-309208-6 Disponível em: https://books.google.co.ao/books?id=071VDwAAQBAJ&pg=PR2 &lpg=PR2&dq=ISBN+978-92-5-309208-6&source=bl&ots=ihFeG0H0S6&sig=THz9dCBLdN7IE0Z93eliG Kk4C0w&hl=pt-PT&sa=X&ved=2ahUKEwij6ePNuqzdAhUvxoUKHazCAWkQ6AE wAXoECAkQAQ#v=onepage&q=ISBN%20978-92-5-309208- 6&f=false. Last accessed on July 13, 2018.

33. Faria, C. (2010). Ciclo do Carbono. Infoescola, INFOESCOLA, Brazil. Available at: http://www.infoescola.com/biologia/ciclo- do-carbono/

34. Fernandes, A. M. (2014). Analysis of wood production for the sustenance of domestic energy to urban centers in Mozambique. Thesis: Universidade Federal do Paraná.

35. Fischer, M., & Trujillo, W. (1999). Carbon fixation in tropical pastures in neotropical acid soil savannas. International Seminar on Livestock Intensification in Central America: economic and environmental benefits. Costa Rica. 135p.

36. Francisco, E. J., Africano, C. G., Sanfilippo, M., Quintana, Y. G., Martínez, I. C., & Crespo, Y. A. (2014). Structure and composition of the Miombo forest of the northern sector of Canjombe, Angola. Revista Forestal Baracoa vol. 33, Special Issue 2014. Scientific article, pp. 306-316. ISSN: 2078-7235.

37. Frondizi, I. M. de R. L. (2009). The Clean Development Mechanism: Guidance Guide. Rio de Janeiro: Imperial Novo Milênio.

38. Frost, P. (1996). The Ecology of Miombo Woodlands. In: Campbell, B. The Miombo in transition Woodland and Welfare in Africa. South Africa. p 19 -39.

39. GREENPEACE. (2007), What to do to save the forest. Brazil. Available at: http://www.greenpeace.org/brasil/amazonia. Date of Access: 08/11/2017

40. Henriques, I. C. M., Monteiro, A. & Moreira, I. (2009). Effect of phytosanitary treatments on the yield of potato (Solanum tuberosum L.) cultivars in the Huambo Plateau (Angola). Revista de Ciências Agrárias 32, p.182-193.

41. Henson, R. (2009). Climate Change - Symptoms, Science, Solutions. Civilização Editores, Porto. Pp. 384. ISBN 978- 989550-725-2.

4 2.IDF - Instituto de Desevolvimento Florestal. (2008). Republic of Angola. Ministério da Agricultura. Programa De Desenvolvimento E Gestão Para O Sector Florestal.

43. IDF. (2004). Legislação Vigente, Fauna e Flora. Compilation Ed. Pp 231.

44. IEE. (2008). Institute of Electronics and Energy- national biomass reference center. Charcoal technical, social, environmental and economic aspects 2-8p.

45. IPCC - Intergovernmental Panel On Climate Change. (1996c): Climate Change 1995, Economic and Social Dimension of Climate Change. Contribution of Working Group III to the Second Assessment of the IPCC. Eds. JP Bruce, H. Lee, EF Haites, Cambridge University Press, Reino Unido.

46. IPCC. (2006). Guidelines for national greenhouse gas inventories. Geneva: IPCC, 2006. Available at: http://www.ipcc-nggip.iges.or.jp/public/2006gl/index.html (Short URL: http://goo.gl/6zlcgd) Access Date: 25 Sept. 2018.

47. IPCC. (2007). Climate change 2007: mitigation of climate change. Contribution of Working Group III to the Fourth Assessment Report of the IPCC. Cambridge University Press, Cambridge, UK.

48. IPCC. (2007b). Climate Change 2007: The Physical Science Basis. (Available at: http://www.ipcc.ch/). Access Date: 09/14/2018

49. IPCC. 2007b. Climate Change 2007: The Physical Science Basis. (Available at: http://www.ipcc.ch/). Access Date: 09/14/2018

50. IPCC. (2013). IPCC: Guidelines for national greenhouse gas inventories. IPCC, Switzerland, 2013.

51. Johnson, I. & Coburn, R. (2010). Trees for carbon sequestration, Prime facts for profitable, adaptive and sustainable primary industries, N° 981, State of New South Wales through Department of Industry and Investment (Industry & Investment NSW).

52. Kaniama, R. (2017). My Mother's Goat: the secret of wealth: ISTC, 2017. 44p. ISBN: 978-989-20-7398-9.

53. Lombardi, A. (2008). Carbon Credits and Sustainability. São Paulo: Lazuli. Manfrinato, Warwick (org.). Areas of permanent preservation and legal reserve in the context of climate change mitigation: climate change, the forest code, the Kyoto Protocol and the clean development mechanism. Rio de Janeiro: The Nature Conservancy, 2005.

54. Lopes, R. B., & Miola, D. T. B. (2010). Sequestro de carbono em diferentes fitofisionomias do Cerrado, Revista Digital FAPAM, Vol. 2, (2), pp. 127-143.

55. Manuel, M., Rui, P. R., António, G. N. (2015). Forest covers and the recovery of fertility of Ferralitic Soils of Angola. Journal of Agricultural Sciences. 38. 598-611. 10.19084/RCA15142.

56. Martins, F.R. (1991). Estrutura de uma floresta mesófila. Campinas: Unicamp.

57. MEA - Ministry of Energy and Water. (2016). Environmental and Social Management Framework (QGAS) for the Project for the Institutional Development of the Water Sector (PDISA), Angola. National water directorate, project coordination unit. SFG2527, 08/2016.

58. Meister, K., Ashton, M. S., Craven, D., & Griscom, H. (2012). Carbon dynamics of tropical forests. In: Ashton, M.; Spalding, T. D.; Gentry, B. M. S. Managing Forest Carbon in a Changing Climate. Berlin: Springer. 51-75.

59. MINADERP. (2010). Ministry of Agriculture, Rural Development and Fisheries. Politica Nacional de Florestas, Fauna Selvagem e Áreas de Conservação. Diário da Répública la Série n.° 8. Luanda.

60. MINAGRI. (2011). Forest Profile of Angola. Enough.

61. MINPLAN. (2010). Ministry of Planning of the Republic of Angola. Report on the Millennium Development Goals. 64-72 p.

62. Mittermeier, R.A., Myers, N., Gil, P.R., & Mittermeier, C.G. (2000). Hotspots; the Earth's Biologically Richest and Most Endangered Terrestrial Ecoregions. Washington DC, CEMEX and Conservation International

63. Ndendica, D. A. (2014). Elaboration of the Strategic Plan for the Relaunch of the Experimental Station of Sacaála. Bachelor's Dissertation in Forest Engineering, Faculty of Agricultural Sciences, José Eduardo dos Santos University, Huambo, Angola.

10-11p.

64. Neto, E. C. (2011). Tons on shoulders. São Paulo: Schoba.

65. Nilsson, K., et al. (2011). Forests, trees, and human health. Springer. Gavle-Sweden. 409 pp. ISBN 978-90-481-9805-4.

66. Nogueira, M. D. (1970). Carta de Solos do Centro de Estudos da Chianga. In: IIAA - Instituto de Investigação Agronómica de Angola. Série Científica. No. 14. Nova Lisboa, Angola. 72 pp.

67. Ohse, S. et al. (2007). Review: Carbon Sequestration by Microalgae and Forests and the Ability of Microalgae to Produce Lipids. Series No. 36. Florianópolis. 39-74p.

68. UN- United Nations Organization. (2000). FRA 2000: report on the state of the world's forests to the year 2000.p5.

69. Palácios, G., Lara-Gomez, M., Márquez, A., Vaca, J. L., Ariza, D. Lacerda, V., & Navarro-Cerrillo, R. M. (2015). Monitoring deforestation in Huambo Province using detection technologies and geographic information systems. SASSCAL Project Proceedinds. Huambo, Angola. 182 pp.

70. Pedro, M. S. (2010). Balanço de Carbono no Sector do Pinheiro

Bravo from the National Forest of Leiria, Aveiro - Portugal. P.12 Available at   :

< http://ria.ua.pt/bitstream/10773/4272/1/Disserta%C3%A7%C3% A3o.pdf>. Access
Date: 08/09/2018.

71. UNDP. United Nations Development Programme in Angola. (2015). COP21
2015 Special Weekly Press Review. No. 80 - November 25 to December 12,
2015. 2 p.

72. Queiroz, W.T. (2012). Amostragem em Inventário Florestal. Belém:
Universidade Federal Rural da Amazônia. 441 p. ISBN 978-85-7295-070-1

73. Ramos, M.C.P. (2012). "Environment, Education and Interculturality." Journal
Times and Spaces in Education, 8, p.27-40.

74. Raven, P. H., Evert, R. F., & Eichhom, S. E. (2001). Plant biology. (6a ed).
Publisher: Guanabara Koogan. 906 p.

75. Reis, M. G. F., et al. (1994). Sequestro e armazenamento de carbono em
florestas nativas e plantadas no Estados de Minas Gerais e Espírito Santo. In:
Seminário Emissão x Sequestro de CO2: Uma oportunidade de negócios para
o Brasil,: Rio de Janeiro. Annals. Rio de Janeiro: CVRD, 1994. p. 155-195.

76. Renner, R. M. (2004). Sequestro de Carbono e a Viabilização de Novos
Reflorestamentos no Brasil. Curitiba.132p.

77. Rhett, A. B. (2009). Brazil's plan to save the Amazon forest. Mongabay.
Translated by Marcela V.M. Mendes, 2010.

78. Ribeiro, N., Sitoe, A. A., Guedes, B. S., & Staiss, C. (2002). Manual de
Silvicultura Tropical. Universidade Eduardo Mondlane, Faculty of Agronomy
and Forest Engineering, Department of Forest Engineering. Maputo, 2002.
FAO, Project GCP/Moz/056/Net.

79. Ribeiro, T. (1994). O Jardim Comum Europeu: Novos Desafios Ambientais,
Queztal Editores/F.L.A.D, Lisbon (Portugal).

80. Rua, T.A. (2013). Project: Environmental Refugees. Climate change and forced
migration". Catholic University of Peru.

81. Saatchi et al. (2011). Benchmark map of forest carbon stocks in tropical regions
across three continents. Proceedings of the National Academy of Sciences of
the USA 108(24).

82. Sanfilippo, M. (2014). Thirty Angolan miombo trees. Field guide for identification.
COSPE, Firenze.

83. Schumacher, M.V. (1996). Nutrient cycling as a basis for sustained production in forest ecosystems. In: Simpósio sobre Ecossistemas Naturais do Mercosul: O Ambiente da Floresta, 1. Santa Maria. Proceedings. Santa Maria: UFSM/CEPEF, 6577p.

84. Seiffert, M. E. B. (2009). Mercado de Carbono e Protocolo de Kyoto. São Paulo: Atlas.

85. Silva, J. M. S. (2015). Produção e distribuição de biomassa em clones de Eucalyptus grandis x Eucalyptus urophylla no Município de Macaíba-RN. 2015. 37f. Dissertation (Master in Forest Sciences). Federal University of Rio Grande do Norte.

86. Silveira, P., H. S. Koehler, C. R. Sanquetta & J. E. Arce. (2008). The State of the Art in estimating biomass and carbon in forest formations, Floresta PR, Vol. 38, (1), pp.185-206.

87. Sitoe, A., Guedes, B., & Nhantumbo, I. (2013). Baseline, Monitoring, Reporting and Verification for REDD+ in Mozambique. Country report. IIED, London. ISBN 978-178431-044-8.

88. Soares, C. P. B., & Oliveira, M. L. R. (2002). Equações para estimar a quantidade de carbono na parte aérea de arvores de eucalipto em Viçosa, Minas Gerais. Revista Árvore, Viçosa-MG, v.26, (5), p.533-539.

89. Soares, S. C., & Motta, A. L. S. (2010). Decrease of Natural Forests in the World. Available at: http://www.excelenciaemgestao.org/Portals/2/documents/cneg6/ proceedings/T10_0244_1316.pdf.

90. Stern, A. (1977). Air pollution: the effects of air pollution, Vol II. 3rd edition. United Kingdom: Academic Press, Inc.

91. Teixeira, D. et al. (2010). Carbon credit market.

Available   at

< http://www.infobibos.com/Artigos/2010_2/CreditoCarbono/index .htm> Date Accessed: 09/14/2018.

92. Tomé, J. A. P. M. (2014). The quantification of carbon stored by Portuguese forests. Lisboa: Centro de Estudos Florestais Forchange.

93. UNFCCC. (1999). In: The Marrakesh Accords Version: 10-11-01; 5:29.

94. Vaccaro, S. (2007). Phytosociological characterization of three successional stages of a deciduous seasonal forest, in the Municipality of Santa Tereza-RS. PhD Thesis.

95. Van der Werf, G. R., D. C., et al. (2009). CO2 Emissions from Forest Loss. Nature Geoscience 2: 737 - 738. White, 1981.

96. Victorino, A. (2012). Forests for people: Proceedings of the International Year of Forests 2011. Alexander von Humboldt Biological Resources Research Institute and Ministry of Environment and Sustainable Development. Bogotá, D.C., Colombia. 2427p. ISBN: 978-958-8343-73-0 Available at: http://repository.humboldt.org.co/bitstream/handle/20.500.11761/ 31369/230.pdf?sequence=1&isAllowed=y. Date de Acesso: 13/07/ 2018.

97. Vieira, G., Sanquetta, C. R., Kluppel, M. L. W., & Barbeiro, L. S. S. (2009). Carbon contents in plant species from caatinga and cerrado title. Rev. Acad., Ciênc. Agrár. Ambient., Curitiba, v. 7, (2), p. 145-155, abr./jun.

98. Wagner, T., Beirle, S., Grzegorski, M., Sanghavi, S., & Platt, U. (2004). Global long term data sets of the atmospheric H2O column derived from Gome and Aciamachy - anomalies during the strong

El-Nino event 1997/1998. University of Heidelberg, Germany. Available at: http://joseba.mpch-

mainz.mpg.de/pdf_dateien/esa_salzburg_wagner_04.pdf. Access Date: 09/14/2018

99. Wolf, S. et al. (2011). Strong seasonal variations in net ecosystem CO2 exchange of a tropical pasture and afforestation in Panama. Agricultural and Forest Meteorology, v. 151, p. 1139-1151.

100. World Health Organization. (2018). MDG Goal 7: Ensure

environmental sustainability - Other MDGs. WHO: Regional Office   forAfrica. Disponívelem:

http://www.aho.afro.who.int/profiles_information/index.php/Angol a:MDG_Goal_7:_Ensure_environmental_sustainability_-_Other_MDGs/pt.

101. Zelarayán, M. L. C., et al. (2015). Impact of degradation on the total carbon stock of riparian forests in eastern Amazonia, Brazil. Acta amazónica. Vol. 45, (3) p. 271 - 282.

ANNEXES OR APPENDICES

ANNEXES OR APPENDICES

Appendix I. Carbon Sequestration Calculations by Species

Tree species	DBH (m)	Height(m)	Age (years)	Case a (cm)	Branches (cm)	DCF	Wood Volume	Biomass to air	Root %	Net biomass rate per tree	Annual biomass rate	Carbon sequestration
A. antunesiana	0,1	4,3	15,6	0,5	253,0	6,2	3183,6	3443,2	0,3	4338,5	278,1	139,1
A. bohemii	0,1	5,1	12,6	0,4	162,1	4,6	3009,9	3177,0	0,3	4003,1	317,7	158,9
B. madagascariensis	o,1	4,5	14,5	0,4	200,0	4,9	3882,2	4087,5	0,3	5150,3	355,2	177,6
B. bohemii	0,1	4,8	15,3	0,6	245,7	9,2	3406,0	3661,5	0,3	4613,5	301,5	150,8
B. longuiflora	0,1	4,7	12,8	0,5	212,6	5,5	4804,2	5022,8	0,3	6328,7	494,4	247,2
B. spiciformis	0,2	9,3	18,1	0,7	291,8	8,7	2661,9	2963,2	0,3	3733,6	206,3	103,1
H. acida	0,2	3,8	12,6	0,4	234,2	6,0	971,2	1211,8	0,3	1526,8	121,2	60,6
1. angolensis	0,2	6,1	16,5	0,5	273,5	6,7	1977,4	2258,2	0,3	2845,3	172,4	86,2
Monote spp	0,1	3,5	15,3	0,5	209,1	7,4	4056,9	4273,9	0,3	5385,1	352,0	176,0
O.schweinfurthiana	0,1	3,2	13,0	0,3	172,8	4,6	10815,6	10993,3	0,3	13851,6	1065,5	532,8
P. curatelifolia	0,1	5,4	13,4	0,4	183,3	5,0	5281,5	5470,2	0,3	6892,5	514,4	257,2
P. angolensis	0,1	3,4	12,0	0,3	120,0	4,4	12816,3	12941,1	0,3	16305,7	1358,8	679,4
P. febrifugum	0,1	3,0	12,5	0,6	165,0	3,6	3768,0	3937,1	0,3	4960,8	396,9	198,4
R. engleriana	0,1	3,8	14,0	0,5	235,0	5,2	4239,0	4479,7	0,3	5644,4	403,2	201,6
S. quinquebola	0,1	4,5	12,6	0,4	132,0	0,2	4684,9	4817,5	0,3	6070,0	481,7	240,9
S. guineense	0,1	5,5	14,0	0,4	185,4	4,3	3162,3	3352,4	0,3	4224,0	301,7	150,9
T. brachystemma	0,1	1,7	11,0	0,3	153,8	4,5	4621,7	4780,3	0,3	6023,2	547,6	273,8
U. benguellensis	0,1	3,6	13,0	0,4	151,6	4,2	4138,5	4294,7	0,3	5411,3	416,3	208,1
U. gosswoilleri	0,1	3,8	12,0	0,6	159,4	4,7	3529,3	3694,0	0,3	4654,4	387,9	193,9
E. grandis	0,3	20,3	48,5	2,0	372,6	50,2	1625,8	2050,6	0,3	2583,7	53,3	26,6
P. patula	0,3	21,1	52,0	2,0	349,7	49,8	1681,6	2083,2	0,3	2624,8	50,5	25,2

Appendix II. Carbon Sequestration calculations by parcel

Parcel	DBH (m)	Height(m)	Age (years)	Bark (cm)	Branche s (cm)	DCF	Volum e of Wood	Aerial Biomass	Root %	Net biomass rate per tree	Annual biomass rate	Carbon Sequestration
P1 Miombo	0,12	5,1	15,3	0,6	1,8	2,8	0,019	5,26263622	0,26	6,63092163 9	0,43	0,22
P2 Miombo	0,108	5,5	13,9	0,56	1,4	2,7	0,017	4,63065548	0,26	5,83462590	0,42	0,21
P3 Miombo	0,133	5,75	14,5	0,4	2	1,9	0,026	4,32209645	0,26	5,44584152	0,38	0,19
P4 Miombo	0,134	6	16	0,28	2,9	1,3	0,028	4,53493745	0,26	5,71402118	0,36	0,18
P5 Miombo	0,141	5,7	15	0,51	3,6	2,4	0,029	6,55643255	0,26	8,26110502	0,55	0,28
P6 Miombo	0,12	5,6	14,9	0,44	1,8	2,1	0,021	4,34620849 3	0,26	5,47622270	0,37	0,18
P1Brito	0,29	21	52	2,3	4,65	6,8	0,458	14,1864032	0,26	17,8748681	0,34	0,17
P2Brito	0,304	21,6	52	2,1	4,3	6,2	0,517	13,1065884	0,26	16,5143014	0,32	0,16
P3Brito	0,27	19,7	44,4	1,8	2	5,3	0,372	9,47725883	0,26	11,9413461	0,27	0,13
P4Brito	0,29	20,9	52	2,25	2,74	6,6	0,456	12,0768613	0,26	15,2168452	0,29	0,15
P5Brito	0,35	20,3	49,86	1,9	4,85	5,6	0,645	12,9942850	0,26	16,3727991	0,33	0,16

Figure 13. Photographic record of the Brito Teixeira Exotic Forest (Chianga, Huambo)

Figure 14. Photographic record of the Miombo Native Forest (Chianga, Huambo)

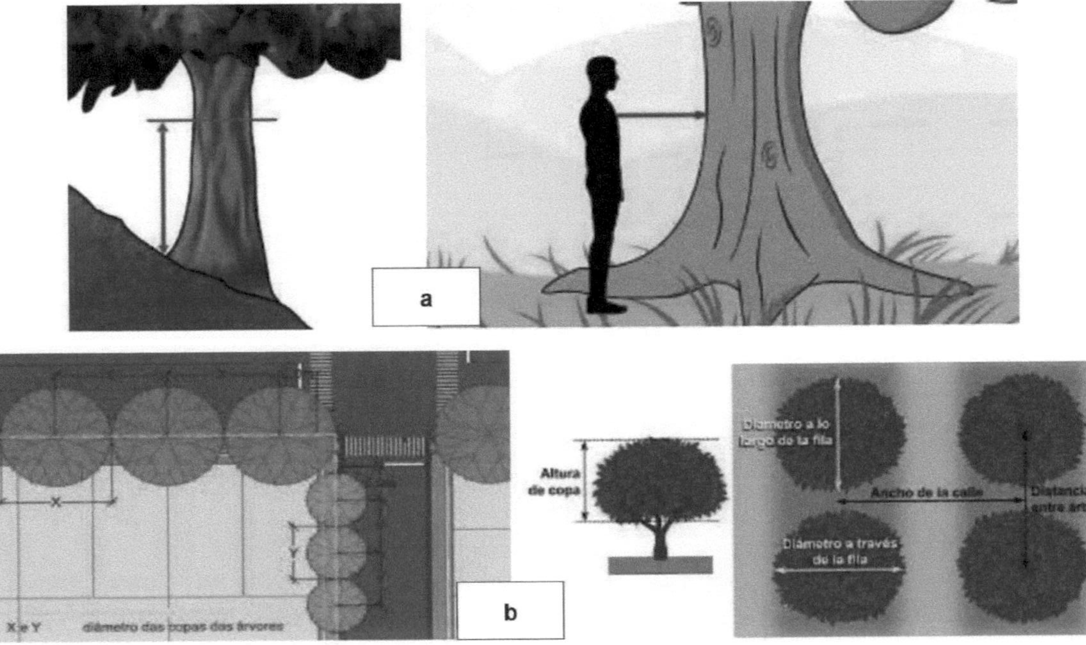

Figure 15. Illustration of measurement methods for diameter at breast height (a) and crown diameter (b)

I **want** morebooks!

Buy your books fast and straightforward online - at one of world's fastest growing online book stores! Environmentally sound due to Print-on-Demand technologies.

Buy your books online at
www.morebooks.shop

Kaufen Sie Ihre Bücher schnell und unkompliziert online – auf einer der am schnellsten wachsenden Buchhandelsplattformen weltweit! Dank Print-On-Demand umwelt- und ressourcenschonend produziert.

Bücher schneller online kaufen
www.morebooks.shop